FA-240/December 2002

A Needs Assessment of the U.S. Fire Service

A Cooperative Study
Authorized by U.S. Public Law 106-398

ACKNOWLEDGEMENTS

NFPA Project Manager: John R. Hall, Jr., Ph.D.
NFPA Senior Statistician: Michael J. Karter, Jr.
USFA Project Officer: Mark A. Whitney

It required literally a cast of thousands to make this study of the needs of the US fire service possible. First and foremost, we wish to thank the more than 12,000 US fire departments who carefully reviewed their departments' capabilities and described those capabilities in forms submitted to us for use in this study. The response has been literally unprecedented in the history of studies of the US fire service, and we expect that we will all be learning lessons far into the future.

Second, we want to thank the many individuals who guided us in selecting the most important questions to ask and the most appropriate interpretations of answers received. These include our Technical Advisory Group:

- Steve Coffman, Captain, Dallas (TX) Fire Department
- Arthur Cota, Director, California Fire Service Training
- Robert DiPoli, Chief, Needham (MA) Fire Department
- Jeff Dyar, U.S. Fire Administration
- Dr. James Genovese, US Army Soldier and Biological Chemical Command, Aberdeen Proving Grounds
- Joseph Kay, Battalion Chief, Dallas (TX) Fire Department
- Eric Lamar, International Association of Fire Fighters
- Edward Plaugher, Chief, Arlington County (VA) Fire Department
- Ernest Russell, State Fire Marshal, Illinois
- Gary Santoro, Fire Marshal, Wethersfield (CT) Fire Department
- Heather Schafer, Executive Director, National Volunteer Fire Council
- Eric Tolbert, formerly Administrator, North Carolina Emergency Management, and currently on staff with FEMA
- Jeff Wagoner, Campbell County (WY) Fire Department
- Mark A. Whitney, Fire Programs Specialist, U.S. Fire Administration

We also received extensive and essential comments at several stages from colleagues at NFPA:

- Gary Tokle, Assistant Vice President, Public Fire Protection Division
- Carl Peterson, Assistant Director, Public Fire Protection Division
- Steven Foley, Senior Fire Service Specialist, Public Fire Protection Division
- Bruce Teele, Senior Fire Service Specialist, Public Fire Protection Division
- Rita Fahy, Manager – Fire Data Bases and Systems, Fire Analysis & Research Division

We also received extensive and essential comments at several stages from the staff at USFA and we greatly appreciate their insights.

Lastly, we want to thank the administrative personnel at NFPA, whose painstaking attention to detail and extended hours of work were instrumental in transforming a set of questions and a stack of forms into a unique database and this analysis report:

- John Baldi
- John Conlon
- Frank Deely
- Myles O'Malley
- Kevin Tape

- Norma Candeloro
- Helen Columbo
- Laurie Eisenhauer

EXECUTIVE SUMMARY

PL 106-398, Section 1701, Sec. 33 (b) required that the Director of the Federal Emergency Management Agency (FEMA) conduct a study in conjunction with the National Fire Protection Association (NFPA) to
  (a) define the current role and activities associated with the fire services;
  (b) determine the adequacy of current levels of funding; and
  (c) provide a needs assessment to identify shortfalls.

The Fire Service Needs Assessment Survey was conducted as a census, with appropriate adjustments for non-response. The NFPA used its own list of local fire departments as the mailing list and sampling frame of all fire departments in the US. In all, 26,354 fire departments were mailed survey forms.

The content of the survey was developed by NFPA, in collaboration with an ad hoc technical advisory group consisting of representatives of the full spectrum of national organizations and related disciplines associated with the management of fire and related hazards and risks in the U.S. Overall, NFPA received 12,240 completed surveys and has edited, coded, and keyed 8,416 surveys for analysis in this report. The overall response rate is 46%, which is unusually high for a survey involving a large number of smaller departments.

Because NFPA prepared two preliminary reports based on the first 5,100 surveys keyed and those results are very similar to the results based on 8,416 surveys, the authors believe that the surveys keyed late and so not included in this analysis would not, if analyzed, materially affect the results, either nationally or by community size. In particular, all surveys from departments protecting populations of 50,000 population or more were keyed for this analysis, and a sufficient number of surveys from each of the population intervals for smaller communities have also been keyed to assure a statistically valid sample. However, the additional surveys keyed will permit a much larger share of US fire departments to have participated, in what clearly is shaping up as the highest-participation and most-detailed database on fire service resources and needs ever assembled.

The US Fire Service – Revenues and Budgets

- Most of the revenues for all- or mostly-volunteer fire departments come from taxes, either a special fire district tax or some other tax, including an average of 63-67% of revenues covered for communities of less than 5,000 population.

- Other governmental payments – including reimbursements on a per-call basis, other local government payments, and state government payments – contributed an average of 13% of revenues for communities under 5,000 population.

- Fund-raising contributed an average of 19% of revenues for communities of less than 2,500 population.

- Used vehicles accounted for an average of 42% of apparatus purchased by or donated to departments protecting communities with less than 2,500 population.

- Converted vehicles accounted for an average of 16% of apparatus used by departments protecting communities with less than 2,500 population.

Personnel and Their Capabilities

- There are just over a million active firefighters in the US, of which just over three-fourths are volunteer firefighters. Nearly half the volunteers serve in communities with less than 2,500 population.

Table ES-1. Number of Career, Volunteer, and Total Firefighters by Size of Community

Population Protected	Career Firefighters	Volunteer Firefighters	Total Firefighters
1,000,000 or more	32,700	150	32,850
500,000 to 999,999	28,400	4,900	33,300
250,000 to 499,999	26,600	4,250	30,850
100,000 to 249,999	39,750	8,550	48,300
50,000 to 99,999	37,750	11,000	48,750
25,000 to 49,999	40,000	29,300	69,300
10,000 to 24,999	38,850	86,050	124,900
5,000 to 9,999	12,200	112,300	124,500
2,500 to 4,999	5,050	157,600	162,650
Under 2,500	4,800	408,750	413,550
Total	266,100	822,850	1,088,950

- In communities with less than 2,500 population, 21% of fire departments, nearly all of them all- or mostly-volunteer departments, deliver an average of 4 or fewer volunteer firefighters to a mid-day house fire. Because these departments average only one career firefighter per department, it is likely that most of these departments often fail to deliver the minimum of 4 firefighters needed to safely initiate an interior attack on such a fire.

- An estimated 73,000 firefighters serve in fire departments that protect communities of at least 50,000 population and have fewer than 4 career firefighters assigned to first-due engine companies. It is likely that,for many of these departments, the first arriving complement of firefighters often falls short of the minimum of 4 firefighters needed to safely initiate an interior attack on a structure fire, thereby requiring the first-arriving firefighters to wait until the rest of the first-alarm responders arrive.

- An estimated 233,000 firefighters, most of them volunteers serving in communities with less than 2,500 population, are involved in structural firefighting but lack formal training in those duties.

- An estimated 153,000 firefighters, most of them volunteers serving in communities with less than 2,500 population, are involved in structural firefighting but lack certification in those duties.

- An estimated 27% of fire department personnel involved in delivering emergency medical services (EMS) lack formal training in those duties, most of them serving in communities with less than 10,000 population.

- The majority of fire departments do not have all their personnel involved in emergency medical services (EMS) certified to the level of Basic Life Support and almost no departments have all those personnel certified to the level of Advanced Life Support.

- An estimated 40% of fire department personnel involved in hazardous material response lack formal training in those duties, most of them serving in smaller communities.

- More than four out of five fire departments do not have all their personnel involved in hazardous material response certified to the Operational level and almost no departments have all those personnel certified to the Technician level.

- An estimated 41% of fire department personnel involved in wildland firefighting lack formal training in those duties, with substantial needs in all sizes of communities.

- An estimated 53% of fire department personnel involved in technical rescue service lack formal training in those duties. In every population group of communities with less than 500,000 population, at least 40% of fire department personnel involved in technical rescue service lack formal training in those duties.

- An estimated 792,000 firefighters serve in fire departments with no program to maintain basic firefighter fitness and health, most of them volunteers serving communities with less than 5,000 population.

Fire Prevention and Code Enforcement

- An estimated 83.9 million people (29% of the US resident population in 2001) are protected by fire departments that do not provide plans review, an estimated 128.8 million (45%) by departments that do not provide permit approval, and an estimated 140.5 million (49%) by departments that do not provide routine testing of active systems (e.g., fire sprinklers). Each of these services may be provided by another agency or organization in these communities.

- An estimated 120.1 million people (42%) are protected by fire departments that do not have a program for free distribution of home smoke alarms.

- An estimated 136.3 million people (48%) are protected by fire departments that do not have a juvenile firesetter program.

- An estimated 78.0 million people (27%) are protected by fire departments that do not have a school fire safety education program based on a national model curriculum. Moreover, independent data on the breadth of implementation of such curricula indicate that most fire departments reporting programs provide only annual or occasional presentations based on material from such a curriculum.

- An estimated 20.9 million people (7%) live in communities where no one conducts fire-code inspections. Two-fifths of this population live in rural communities, with less than 2,500 population.

Facilities, Apparatus and Equipment

- Roughly 15,500 fire stations (32% of the estimated 48,500 total fire stations) are estimated to be at least 40 years old, roughly 27,500 fire stations (57%) have no backup power, and nearly 38,000 fire stations (78%) are not equipped for exhaust emission control.

- Using maximum response distance guidelines from the Insurance Services Office and simple models of response distance as a function of community area and number of fire stations, developed by the Rand Corporation, it is estimated that three-fifths to three-fourths of fire departments have too few fire stations to meet the guidelines.

- Just over 13,000 fire engines (pumpers) (16% of all engines) are 15 to 19 years old, another 17,000 (21%) are 20 to 29 years old, and just over 10,000 (13%) are at least 30 years old. Therefore, half of all engines are at least 15 years old.

- Among fire departments protecting communities with less than 10,000 population, at least 10% of departments are estimated to have no ladder/aerial apparatus but to have at least one building 4 stories high or higher in the community.

- Overall, fire departments do not have enough portable radios to equip more than about half of the emergency responders on a shift. The percentage of departments that cannot provide radios to all emergency responders on a shift is especially high for communities under 2,500 population.

- The majority of fire department portable radios are not water-resistant, and more than three-fifths lack intrinsic safety in an explosive atmosphere. These needs are more pronounced in smaller communities.

- An estimated one-third of firefighters per shift are not equipped with self-contained breathing apparatus (SCBA). Nearly half of SCBA units are at least 10 years old.

- Nearly half of the emergency responders per shift are not equipped with personal alert system (PASS) devices.

- An estimated 57,000 firefighters lack personal protective clothing, most in departments protecting communities with less than 2,500 population. An estimated one-third of personal protective clothing is at least 10 years old.

Communications and Communications Equipment

- Three-fifths to four-fifths of fire departments (62-82%, by size of community protected) say they can communicate at incident scenes with their Federal, state, and local partners. Of these, though, only two-fifths say they can communicate with all their partners. This means only about one-fourth of departments overall can communicate with all partners.

- Nearly half of all fire departments have no map coordinate system. Most departments with a map coordinate system have only a local system. Interoperability of spatial-based information systems, equipment, and procedures probably will not be possible under these circumstances, for multiple jurisdiction/ agency catastrophic disaster response.

- One-fourth of departments (one-third of rural fire departments) have 911-Basic for telephone communication. Two-thirds have 911-Enhanced, and 6% have no special 3-digit number.

- Overall, one community in 12 (9%) has primary responsibility for dispatch operations lodged with the fire department, but that fraction rises to four-fifths for communities of at least 1 million population.

- One-third of communities have primary dispatch responsibility lodged with the police department, and another one-third with a combined public safety department.

- Two-fifths of departments lack a backup dispatch facility, including nearly half of departments protecting communities with less than 2,500 population.

- Two-fifths of departments lack Internet access.

Ability to Handle Unusually Challenging Incidents

- Only 11% of fire departments can handle a <u>technical rescue with EMS at a structural collapse of a building with 50 occupants</u> with local trained personnel.

  ➢ Nearly half of all departments consider such an incident outside their scope.

  ➢ Only 11% can handle the incident with local specialized equipment.

  ➢ Only 19% have a written agreement to direct use of non-local resources.

  ➢ All needs are greater for smaller communities.

- Only 13% of fire departments can handle a <u>hazmat and EMS incident involving chemical/biological agents and 10 injuries</u> with local trained personnel.

  ➢ Two-fifths of all departments consider such an incident outside their scope.

  ➢ Only 11% can handle the incident with local specialized equipment.

  ➢ Only 21% have a written agreement to direct use of non-local resources.

  ➢ All needs are greater for smaller communities.

- Only 26% of fire departments can handle a <u>wildland/urban interface fire affecting 500 acres</u> with local trained personnel.

  ➢ One-third of all departments consider such an incident outside their scope.

  ➢ Only 22% can handle the incident with local specialized equipment.

  ➢ Nearly half the departments that consider such an incident within their scope, and 33% overall, have a written agreement to direct use of non-local resources.

  ➢ All needs for local resources are greater for communities of 5,000 to 249,999 population, and the need for written agreements is greater for smaller communities.

- Only 12% of fire departments can handle <u>mitigation of a developing major flood</u> with local trained personnel.

  ➢ The majority of departments consider such an incident outside their scope.

  ➢ Only 11% can handle the incident with local specialized equipment.

> Only 13% have a written agreement to direct use of non-local resources.

> All needs are greater for smaller communities.

New and Emerging Technology

- One-fourth of fire department now own thermal imaging cameras, but most that do not have them now have no plans to acquire them.

- Only one department in 28 has mobile data terminals, only one in 50 has advanced personnel location equipment, and only one in 23 has equipment to collect chemical or biological samples for remote analysis. Most departments have no plans to acquire them.

TABLE OF CONTENTS

LIST OF TABLES AND FIGURES

LIST OF TABLES AND FIGURES (Continued)

LIST OF TABLES AND FIGURES (Continued)

INTRODUCTION

FEMA Survey Project on Needs of the US Fire Service

The report that follows presents results based on data from US local fire departments participating in a needs assessment survey.

Public Law 106-398, Fire Investment and Response Enhancement (FIRE) Act, Title XVII – Assistance to Firefighters, recognized that America's fire departments provide service and protection with impact far beyond the borders of the communities that support them. In order to provide this service and protection with the effectiveness, speed, and safety that their home communities and the nation as a whole demand, many fire departments will need to increase their resources, in any of several categories.

PL 106-398 created a fund to support worthy proposals to address these needs. But PL 106-398 also recognized that our current understanding of the magnitude and nature of fire department needs is not well defined. Furthermore, the rationale for Federal government assistance to meet these needs is also in need of greater definition, given the normal presumption that routine fire protection is a local function, set to meet locally defined goals and supported by local resources.

Accordingly, PL 106-398, Section 1701, Sec. 33 (b) required that the Director of the Federal Emergency Management Agency (FEMA) conduct a study in conjunction with the National Fire Protection Association (NFPA) to
  (d) define the current role and activities associated with the fire services;
  (e) determine the adequacy of current levels of funding; and
  (f) provide a needs assessment to identify shortfalls.

The Act identifies several categories of types of resources:

  A. Firefighting personnel
  B. Training for firefighting personnel
  C. Rapid intervention teams
  D. Certification for fire inspectors
  E. Wellness and fitness programs for firefighting personnel
  F. Emergency medical service programs provided by fire departments
  G. Firefighting vehicles
  H. Firefighting equipment, including communications and monitoring
  I. Personal protective equipment for firefighting personnel
  J. Modifications to stations and facilities for reasons of firefighter health and safety
  K. Enforcement of fire codes
  L. Fire prevention programs
  M. Education of the public about arson prevention and detection
  N. Recruitment and retention incentives for volunteer firefighters in fire departments

See Appendix 1 for a more detailed discussion of the statistical methodology used.

The questionnaire principally involved multiple approaches to answering the question "what does the fire department need?". Most of the questions were intended to determine what fire departments have, in a form that could be compared to existing standards or formulas that set out what fire departments should have. Some of the questions asked what fire departments have with respect to certain cutting-edge technologies for which no standards yet exist and no determinations of need have yet been proposed.

The questionnaire also sought to define the emergency-response tasks that fire departments considered to be within their scope. For such tasks the survey asked how far departments would have to go to obtain the resources necessary to address those tasks or an illustrative incident of that type. Clearly, if departments believe the resources they would need are only available from sources separated from them by great distance – and the associated likelihood of significant delay in attaining those resources, then there may be a need for planning, training, or arrangements for equipment that can be more quickly accessed and deployed, to assure timely and effective response.

See Appendix 2 for a copy of the questionnaire.

Glossary

Here are standard definitions for some of the specialized terms used in this report:

Advanced Life Support. Functional provision of advanced airway management, including intubation, advanced cardiac monitoring, manual defibrillation, establishment and maintenance of intravenous access, and drug therapy. [from NFPA 1710, *Standard for the Organization and Deployment of Fire Suppression Operations, Emergency Medical Operations, and Special Operations to the Public by Career Fire Departments*, 2001 edition.]

Basic Life Support. Functional provision of patient assessment, including basic airway management; oxygen therapy; stabilization of spinal, musculo skeletal, soft tissue, and shock injuries; stabilization of bleeding; and stabilization and intervention for sudden illness, poisoning and heat/cold injuries, childbirth, CPR, and automatic external defibrillator (AED) capability. [from NFPA 1710, *Standard for the Organization and Deployment of Fire Suppression Operations, Emergency Medical Operations, and Special Operations to the Public by Career Fire Departments*, 2001 edition.]

Emergency Medical Care. The provision of treatment to patients, including first aid, cardiopulmonary resuscitation (CPR), basic life support (EMT level), advanced life support (Paramedic level), and other medical procedures that occur prior to arrival at a hospital or other health care facility. [from NFPA 1581, *Standard on Fire Department Infection Control Program*, 2000 edition] In this report, reference is made to "EMS" or "emergency medical service," which is the service of providing emergency medical care.

First Responder (EMS). Functional provision of initial assessment (i.e., airway, breathing, and circulatory systems) and basic first-aid intervention, including CPR and automatic external defibrillator (AED) capability. [from NFPA 1710, *Standard for the Organization and Deployment of Fire Suppression Operations, Emergency Medical Operations, and Special Operations to the Public by Career Fire Departments*, 2001 edition.]

Hazardous Material. A substance that presents an unusual danger to persons due to properties of toxicity, chemical reactivity, or decomposition, corrosivity, explosion or detonation, etiological hazards, or similar properties. [from NFPA 1500, *Standard on Fire Department Occupational Safety and Health Program*, 1997 edition.]

Structural Fire Fighting. The activities of rescue, fire suppression, and property conservation in buildings, enclosed structures, aircraft interiors, vehicles, vessels, aircraft, or like properties that are involved in a fire or emergency situation. [from NFPA 1500, *Standard on Fire Department Occupational Safety and Health Program*, 1997 edition.]

Technical Rescue. The application of special knowledge, skills, and equipment to safely resolve unique and/or complex rescue situations. [from NFPA 1670, *Standard on Operations and Training for Technical Rescue Incidents*, 1999 edition.]

Wildland/Urban Interface. The line, area, or zone where structures and other human development meet or intermingle with undeveloped wildland or vegetative fuels. [from NFPA 295, *Standard for Wildfire Control*, 1998 edition]

THE US FIRE SERVICE

Career and Volunteer Fire Departments

Most US fire departments are volunteer fire departments, but most of the US is protected by career firefighters. Tables 1 and 2 (pp. 8-9) provide summary overviews of US fire departments.

Three of every four US fire departments are all-volunteer fire departments, but only one of every four US residents are protected by such a department. Only one in 17 fire departments is all-career, but two of every five US residents is protected by such a department. Fire departments split roughly 9-to-1 between the all- or mostly-volunteer departments vs. the all- or mostly-career departments, but population protected splits roughly 2-to-3 the other way.

Volunteers are concentrated in rural communities, while career firefighters are found disproportionately in large communities. The all- or mostly-career departments account for all of the fire departments protecting communities of at least 1 million population and for more than 90% of the fire departments protecting communities of 250,000 to 999,999 population. All- or mostly-career departments still account for a majority of departments down to communities of at least 25,000 population. Rural communities, defined by the US Bureau of Census as a community with less than 2,500 population, are 99.5% protected by all- or mostly-volunteer departments and account for just over half of all the all- or mostly-volunteer departments in the US.

Community size is related to the US fire service not only in terms of the relative emphasis on career vs. volunteer firefighters but also in terms of the challenges faced by local departments. However, it is possible to exaggerate those differences. Even a rural community can have a large factory complex, a large stadium, or even a high-rise building, with all the technical complexities and potential for high concentration of people or valued property that such a property entails. Even a large city can have a wildland/urban interface region and exposure to the unique fire dangers attendant on such an area. It is likely that every fire department will need to have some familiarity with every type of fire and every type of emergency, if not as part of protecting their own community, then at least in their role as a source of mutual aid or a component of regional or even national response to a major incident.

In any community, fire burns the same way in open or in enclosed spaces. Fire harms people and property in the same ways. And the resources and best practices required to safely address the fire problem – or any other major emergency – tend to be the same everywhere. What may differ is the defined scope of responsibility of the local fire department and the quality and quantity of resources available to the department to perform those responsibilities.

Budgets and Revenue Sources

The first questions of the Needs Assessment Survey focused on big-picture budget and revenue issues. Table 3 (p. 10) asked whether the department has a plan for apparatus replacement on a regular schedule. This is the kind of long-range, capital-budget type of plan that might be more likely in a community with established, institutionalized sources of revenue for the fire department, as one would expect to see in with a career fire department.

Table 3 shows that above a population of 25,000, which is the dividing line for the majority of departments being all- or mostly-career vs. all- or mostly-volunteer, at least 69% of departments in every community-size group have such plans. Below 10,000 population, the majority of departments do not have such plans, and among rural communities, only one department in five has such a plan.

Table 4 (p. 11) addresses the related question of whether the department's normal budget covers the costs of apparatus replacement or whether the department must seek funds in a more ad hoc fashion, such as seeking a special appropriation for such a purchase. Above a population of 50,000, at least 60% of the departments in each population interval cover apparatus replacement in their normal budget. For communities of 25,000 to 49,999 population, half the departments cover apparatus replacement in their normal budget. The percentage with apparatus replacement covered in normal budgets drops to 38% for communities of 10,000 to 24,999 population; to 28% for communities of 5,000 to 9,999 population; to 21% for communities of 2,500 to 4,999 population; and to 12% for communities of less than 2,500 population. Because most departments are small all-volunteer departments serving a rural population, this last figure dominates the results for the US fire service as a whole, where only 22% of departments have apparatus replacement covered by the normal budget.

This result helps to explain why such a large fraction of the initial applications for PL 106-398 were for help in purchasing apparatus. Later questions, having to do with the number of pieces of apparatus that are quite old, will provide additional insight.

The remaining questions in the "Budget Information" section of the survey were asked only of all- or mostly-volunteer fire departments and were designed to further refine the picture of where their revenue comes from and how such departments acquire apparatus. Tables 5 and 6 (pp. 12-13) provide those results. These questions were analyzed only for communities of less than 50,000 population, which is the maximum community size for which at least one-third of departments are all- or mostly-volunteer.

Table 5 shows that most revenues for all- or mostly-volunteer departments are covered by taxes, either a special fire district tax or some other tax. The share of revenues contributed in this way was 72-78% for communities of 5,000 to 49,999 population and 63-67% for communities of less than 5,000 population. Other governmental payments – including reimbursements on a per-call basis, other local government payments, and state government payments – ranged from 9% of revenues for communities of 25,000 to

49,999 population up to 13% of revenues for communities under 5,000 population. Most of the rest was obtained through fund-raising, which ranged from 9% to 10% of revenues contributed for communities of at least 10,000 population up to 19% of revenues contributed for communities of less than 2,500 population.

Table 6 shows that the smaller communities, with less certain sources of revenue, are more likely to obtain their apparatus either used or converted from a non-fire-department design and use. Vehicles that were purchased or, less often, donated used accounted for an average of 9% of apparatus for departments protecting communities with at least 25,000 population but an average of 42% of apparatus for departments protecting communities with less than 2,500 population. Converted vehicles accounted for an average of 4% of apparatus for departments protecting communities with at least 25,000 population but an average of 16% of apparatus for departments protecting communities with less than 2,500 population.

Because converted vehicles were not originally designed for fire department use, it can be especially challenging to assure that they are safe and effective, but it essential that any vehicle, converted or not, be evaluated for its compliance with applicable standards, in order to avoid undue hazard or risk to the firefighters who operate it. A starting point for such an evaluation can be NFPA 1912, *Standard for Fire Apparatus Refurbishing*.

Table 1
Number of Departments and Percent of US Population Protected by Type of Department
(Q. 1, 7, 8)

Type of Department	Number	Percent	Percent of US Population Protected
All Career	1,526	5.8%	40.3%
Mostly Career	1,213	4.6%	18.2%
Mostly Volunteer	3,671	13.9%	15.6%
All Volunteer	19,944	75.7%	25.9%
Total	26,354	100.0%	100.0%

Source: FEMA US Fire Administration 2002
   Survey of the Needs of the US Fire Service

Type of department is broken into four categories. All-career departments are comprised of 100% career firefighters. Mostly-career departments are comprised of 51 to 99% career firefighters, while mostly-volunteer departments are comprised of 1 to 50% career firefighters. All-volunteer departments are comprised of 100% volunteer firefighters.

The above projections are based on 8,027 departments reporting on Questions 1, 7 and 8. Numbers may not add to totals due to rounding.

Q. 1: Population (number of permanent residents) your department has primary responsibility to protect (excluding mutual aid areas)
Q. 7: Total number of full-time (career) uniformed firefighters
Q. 8: Total number of active part-time (call or volunteer) firefighters

Table 2
Department Type, by Community Size
(Q. 1, 7, 8)

Population of Community	All Career		Mostly Career		Mostly Volunteer		All Volunteer		Total	
	Number Depts	Percent	Number Depts	Percent	Number Depts	Percent	Number Depts	Percent	Number Depts	Percent
1,000,000 or more	12	92.3%	1	7.7%	0	0.0%	0	0.0%	13	100.0%
500,000 to 999,999	24	63.2	12	31.6	1	2.6	1	2.6	38	100.0
250,000 to 499,999	42	65.6	18	28.1	2	3.1	2	3.1	64	100.0
100,000 to 249,999	158	73.5	40	18.6	12	5.7	5	2.3	215	100.0
50,000 to 99,999	305	62.6	99	20.4	56	11.5	27	5.5	487	100.0
25,000 to 49,999	389	36.9	288	27.4	226	21.5	149	14.2	1,053	100.0
10,000 to 24,999	438	15.4	493	17.3	1,094	38.5	817	28.7	2,843	100.0
5,000 to 9,999	89	2.4	174	4.8	1,194	32.9	2,172	59.9	3,629	100.0
2,500 to 4,999	30	0.7	54	1.2	629	13.8	3,858	84.4	4,572	100.0
Under 2,500	43	0.3	32	0.2	454	3.4	12,911	96.1	13,440	100.0
Total	1,526	5.8	1,213	4.6	3,671	13.9	19,944	75.7	26,354	100.0

Source: FEMA US Fire Administration 2002
Survey of the Needs of the US Fire Service

Type of department is broken into four categories. All-career departments are comprised of 100% career firefighters. Mostly-career departments are comprised of 51 to 99% career firefighters, while mostly-volunteer departments are comprised of 1 to 50% career firefighters. All-volunteer departments are comprised of 100% volunteer firefighters.

The above projections are based on 8,027 departments reporting on these questions. Numbers may not add to totals due to rounding.

Q. 1: Population (number of permanent residents) your department has primary responsibility to protect (excluding mutual aid areas)
Q. 7: Total number of full-time (career) uniformed firefighters
Q. 8: Total number of active part-time (call or volunteer) firefighters

Table 3
Does Department Have a Plan
for Apparatus Replacement on a Regular Schedule?
by Community Size
(Q. 3)

| Population | Yes | | No | | Total | |
of Community	Number Depts	Percent	Number Depts	Percent	Number Depts	Percent
1,000,000 or more	9	69.2%	4	30.8%	13	100.0%
500,000 to 999,999	36	94.7	2	5.3	38	100.0
250,000 to 499,999	49	76.6	15	23.4	64	100.0
100,000 to 249,999	190	88.4	25	11.6	215	100.0
50,000 to 99,999	391	80.3	96	19.7	487	100.0
25,000 to 49,999	741	70.4	312	29.6	1,053	100.0
10,000 to 24,999	1,719	60.5	1,124	39.5	2,843	100.0
5,000 to 9,999	1,731	47.7	1,898	52.3	3,629	100.0
2,500 to 4,999	1,624	35.5	2,949	64.5	4,572	100.0
Under 2,500	2,779	20.7	10,661	79.3	13,440	100.0
Total	9,269	35.2	17,087	64.8	26,354	100.0

Source: FEMA US Fire Administration 2002
Survey of the Needs of the US Fire Service

The above projections are based on 8,295 departments reporting on Question 3. Numbers may not add to totals due to rounding.

Q. 3: Do you have a plan for apparatus replacement on a regular schedule?

Table 4
Does Department's Normal Budget
Cover the Costs of Apparatus Replacement?
by Community Size
(Q. 4)

| Population | Yes | | No* | | Total | |
of Community	Number Depts	Percent	Number Depts	Percent	Number Depts	Percent
1,000,000 or more	9	69.2%	4	30.8%	13	100.0%
500,000 to 999,999	34	89.5	4	10.5	38	100.0
250,000 to 499,999	42	65.6	22	34.4	64	100.0
100,000 to 249,999	143	66.5	72	33.5	215	100.0
50,000 to 99,999	297	61.0	190	39.0	487	100.0
25,000 to 49,999	527	50.0	526	50.0	1,053	100.0
10,000 to 24,999	1,086	38.2	1,757	61.8	2,843	100.0
5,000 to 9,999	1,028	28.3	2,601	71.7	3,629	100.0
2,500 to 4,999	955	20.9	3,617	79.1	4,572	100.0
Under 2,500	1,609	12.0	11,831	88.0	13,440	100.0
Total	5,730	21.7	20,624	78.3	26,354	100.0

*"No" means the department must raise or seek funds to cover some or all expenses.

Source: FEMA US Fire Administration 2002
        Survey of the Needs of the US Fire Service

The above projections are based on 8,272 departments reporting on Question 4. Numbers may not add to totals due to rounding.

Q. 4: Does your normal budget cover the costs of apparatus replacement?

11

Table 5
For All- or Mostly-Volunteer Departments
Sources of Budget Revenue
by Share (%) of Revenue and Community Size
(Q. 5)

Population of Community	Fire District or Other Tax	Payment per Call	Other Local Payment	State Government	Fund Raising	Other	Total
25,000 to 49,999	78.2%	2.0%	4.9%	1.7%	9.9%	3.3%	100.0%
10,000 to 24,999	76.3	1.8	5.1	3.8	9.4	3.6	100.0
5,000 to 9,999	72.3	1.6	4.4	4.6	13.4	3.7	100.0
2,500 to 4,999	66.7	1.8	5.4	5.4	16.9	3.8	100.0
Under 2,500	62.6	1.9	4.8	6.6	19.1	5.0	100.0

Source: FEMA US Fire Administration 2002
        Survey of the Needs of the US Fire Service

The above projections are based on 5,781 departments reporting on Question 5. Numbers may not add to totals due to rounding.

Q. 5: What share (%) of your budgeted revenue is from [each of the listed alternatives]?

Table 6
For All- or Mostly-Volunteer Departments
Manner of Purchase of Apparatus
by Share (%) of Apparatus and Community Size
(Q. 6)

Population of Community	Purchased New	Donated New	Purchased Used	Donated Used	Converted Vehicles	Other	Total
25,000 to 49,999	87.1%	0.0%	7.8%	1.3%	3.5%	0.3%	100.0%
10,000 to 24,999	82.7	0.6	10.0	1.3	4.9	0.5	100.0
5,000 to 9,999	72.9	0.8	16.2	1.6	8.0	0.5	100.0
2,500 to 4,999	60.5	0.6	23.0	3.4	11.7	0.7	100.0
Under 2,500	39.9	0.7	34.6	7.0	15.8	1.6	100.0

Source: FEMA US Fire Administration 2002
       Survey of the Needs of the US Fire Service

The above projections are based on 5,785 departments reporting on Question 6. Numbers may not add to totals due to rounding.

Q. 6: What share (%) of your apparatus was [each of the listed alternatives]?

14

PERSONNEL AND THEIR CAPABILITIES

Number of Firefighters

Table A indicates the number of career, volunteer, and total firefighters, by the size of the community their fire department protects. These numbers will be used repeatedly throughout the report to convert survey responses phrased in terms of the fraction of a department's firefighters having a characteristic into estimates of the number of firefighters having that characteristic.

Table A. Number of Career, Volunteer, and Total Firefighters by Size of Community
(Q. 1, 7, 8)

Population Protected	Career Firefighters	Volunteer Firefighters	Total Firefighters
1,000,000 or more	32,700	150	32,850
500,000 to 999,999	28,400	4,900	33,300
250,000 to 499,999	26,600	4,250	30,850
100,000 to 249,999	39,750	8,550	48,300
50,000 to 99,999	37,750	11,000	48,750
25,000 to 49,999	40,000	29,300	69,300
10,000 to 24,999	38,850	86,050	124,900
5,000 to 9,999	12,200	112,300	124,500
2,500 to 4,999	5,050	157,600	162,650
Under 2,500	4,800	408,750	413,550
Total	266,100	822,850	1,088,950

The above projections are based on 8,012 departments reporting on Questions 7 and 8. Numbers are estimated to the nearest 50 and may not add to totals due to rounding.

Q. 1: Population (number of permanent residents) your department has primary responsibility to protect (excluding mutual aid areas)
Q. 7: Total number of full-time (career) uniformed firefighters
Q. 8: Total number of active part-time (call or volunteer) firefighters

Table A data on the number of firefighters by community size can be combined with needs-assessment survey results on the percent of firefighters, by community size, who have some need-related characteristic. The result is an estimate of the number of firefighters, by community size, with that need-related characteristic.

Table B indicates the average number of career/paid firefighters per department who are on duty available to respond to emergencies, by size of community the department protects. These figures do not indicate the average number of firefighters per department on duty, because volunteers are not included and every community-size interval has some departments that are not all-career departments.

Table B. Average Number of Career/Paid Firefighters per Department on Duty Available to Respond to Emergencies, by Size of Community (Q. 9)

Population Protected	# of Firefighters
1,000,000 or more	355.1
500,000 to 999,999	217.4
250,000 to 499,999	127.2
100,000 to 249,999	52.9
50,000 to 99,999	24.0
25,000 to 49,999	18.8
10,000 to 24,999	7.3
5,000 to 9,999	3.6
2,500 to 4,999	2.0
Under 2,500	1.0

The above projections are based on 3,177 departments reporting on Question 9.

Q. 9: Average number of career/paid firefighters on duty available to respond to emergencies.

Adequacy of Number of Firefighters Responding

Tables 7-9 (pp. 30-32) provide statistics on numbers of firefighters responding to fight fires under certain circumstances (e.g., as volunteer or career firefighters, to a certain type of fire or with a certain type of apparatus).

These indicators of response profiles can be compared to recently adopted standards regarding the minimum complement of firefighters to permit an interior attack on a structural fire with adequate safeguards for firefighter safety. The comparisons are complicated, however, because most fire departments have both career and volunteer firefighters, while Questions 10-12 asked only about responses by career firefighters alone or volunteer firefighters alone.

Also, in considering the results below, keep in mind that "adequacy" is being assessed here relative to only one of the several objectives of a fire department confronted with a serious fire – the protection of the firefighters themselves from unreasonable risk of injury or death. Relative success in meeting this objective will not necessarily imply anything about the department's ability to reliably achieve the other departmental suppression objectives, whether those be preventing conflagrations, preventing fire from involving an entire large structure, or intervening decisively before the onset of flashover in the room of fire origin. Other analyses will address measures that are more related to those questions.

In addition, success in meeting any of these objectives involves more than a sufficiency of personnel. Equipment of many types is also needed, as are skills and knowledge, as achieved through training and certification. Each of these areas of need is addressed in different parts of the survey.

Volunteer Firefighters

Table 7 provides statistics on the average number of volunteer firefighters who respond to a mid-day house fire, for only the all- or mostly-volunteer fire departments in communities under 50,000 population. Note that a "mostly-volunteer" department might respond with some career firefighters as well, and those numbers are not included in Table 7.

NFPA 1720, *Standard for the Organization and Deployment of Fire Suppression Operations, Emergency Medical Operations, and Special Operations to the Public by Volunteer Fire Departments*, calls for a minimum of 4 firefighters on-site before an interior attack on a structure fire is begun. There are difficulties in applying these standards to Table 7. As noted, responding career firefighters from mostly-volunteer departments are not shown, the statistics shown are average numbers responding rather than minimum numbers responding, and the threshold number of 4 is combined with averages from 3 to 4 in the questionnaire. Nevertheless, some limited observations are possible.

Departments that deliver an average of 1-2 volunteers to a mid-day house fire almost certainly fall below the minimum of 4 firefighters in most responses, at least for departments protecting communities with less than 5,000 population, because Table B indicated that those departments average only 1-2 career firefighters on duty for the department. Departments that deliver an average of 1-2 volunteers (and an unknown number of career firefighters) to a mid-day house fire constituted 3% of departments protecting communities with less than 2,500 population and 3% of departments protecting communities with 2,500 to 4,999 population (see Table 7).

Departments that deliver an average of 3-4 volunteers may fall below the minimum number of 4 firefighters in some responses, particularly in all-volunteer departments and in mostly-volunteer departments protecting communities of less than 2,500 population, where there is, on average, only 1 career firefighter. All-volunteer departments constituted 96% of departments protecting communities with less than 2,500 population, and 84% of departments protecting communities with 2,500 to 4,999 population. The 96% figure for rural communities makes the issue of mostly-volunteer departments in those communities largely moot. Departments that deliver an average of 4 or fewer volunteers to a mid-day house fire constituted 21% of departments protecting communities with less than 2,500 population.

These results suggest that most of the all-volunteer or mostly-volunteer fire departments averaging fewer than 4 firefighters responding to a mid-day house fire, and therefore often failing to achieve the minimum standard response to initiate an interior attack, are departments protecting communities with less than 2,500 population. Because roughly one-fourth of the US resident population live in communities of this size, this suggests roughly 5% of the US population is protected by fire departments that average fewer than 4 firefighters responding to a mid-day house fire and so may often fail to achieve the minimum standard response to initiate an interior attack. (The 5% is calculated as one-fourth of 21%.)

If this is translated into firefighters, then 21% of volunteer firefighters serving communities with less than 2,500 population (from Table A) translates into at least 86,000 volunteer firefighters serving in fire departments where the achievement of a standard minimum response to a mid-day house fire is problematic.

Career Firefighters

Table 8 provides statistics for only the all- or mostly-career fire departments in communities with 10,000 or more population, on the number of career firefighters assigned to an engine or pumper. Note that a "mostly career" department might also respond with some volunteers, and those numbers are not reflected in Table 8. NFPA 1710, *Standard for the Organization and Deployment of Fire Suppression Operations, Emergency Medical Operations, and Special Operations to the Public by Career Fire Departments*, requires a minimum of 4 firefighters on an engine or pumper.

The percentage of departments with fewer than 4 career firefighters assigned to an engine or pumper is 73% for departments protecting 10,000 to 24,999 population, 82% for departments protecting 25,000 to 49,999 population, 76% for departments protecting 50,000 to 99,999 population, 56% for departments protecting 100,000 to 249,999 population, 41% for departments protecting 250,000 to 499,999 population, 40% for departments protecting 500,000 to 999,999 population, and 0% for departments protecting at least a million population.

Because Table A indicates that communities with less than 50,000 population have a substantial number of volunteer firefighters, it is appropriate to focus on departments protecting communities of 50,000 population or more as the ones where the number of responding career firefighters will tend to be the same as the number of responding total firefighters.

This translates into an estimated 73,000 career firefighters serving in fire departments where the community protected has at least 50,000 population and fewer than 4 career firefighters are assigned to an engine. (This figure is calculated as the sum over all community sizes of 50,000 population or more of [number of career firefighters in a community size interval, from Table A] times [percentage of all- or mostly-career fire departments in that interval that assign fewer than 4 people to an engine, from Table 8].)

Table 9 provides statistics comparable to those in Table 8 but for ladder apparatus. There is no comparable simple formula to use in assessing the adequacy of these numbers, so the table is presented without comment.

Extent of Training and Certification, by Type of Duty

Structural Firefighting

Table 10 (p. 33) indicates whether structural firefighting is within the scope of the fire department. Less than 1% of departments say no, nearly all of them in rural communities serving less than 2,500 population.

Table 11 (p. 34) asks how many of the personnel responsible for structural firefighting have received formal training. Answers were solicited in the form of: All, Most, Some, and None. For analysis purposes, "Most" was estimated as 2/3 and "Some" was estimated as 1/3. Based on these assumptions, 233,000 firefighters are estimated to need formal training because they work in departments with responsibility for structural firefighting and have not been so trained. Only 3,000 were in departments protecting communities of 50,000 population or more. These larger communities are the ones where career fire departments dominate. Most of the firefighters estimated to need training were in rural fire departments, protecting communities of less than 2,500 population, and so were almost certainly volunteer firefighters. The breakdown of need by community size, using this approach, is given in Figure 1 as percents and in Table C as numbers of firefighters.

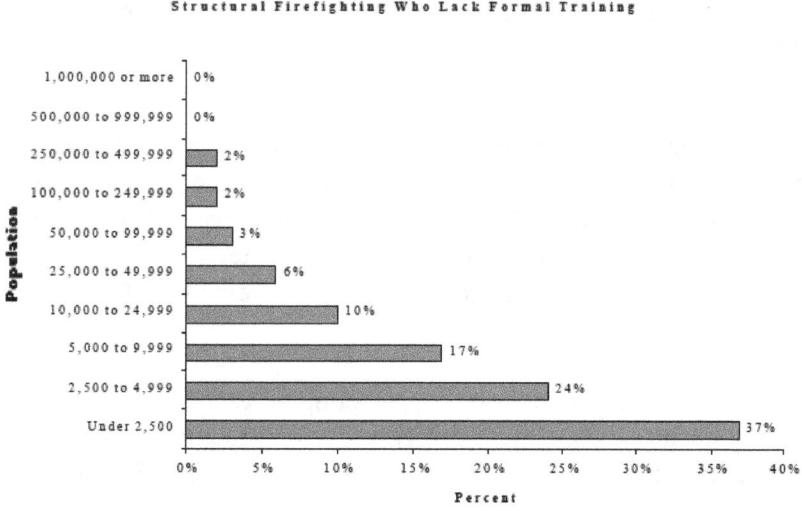

Figure 1. Estimated Percent of Firefighters Involved in Structural Firefighting Who Lack Formal Training

19

**Table C. Estimated Number of Firefighters Involved in
Structural Firefighting Who Lack Formal Training
by Size of Community Protected (Q. 13b)**

Population Protected	Estimated Firefighters Lacking Formal Training
1,000,000 or more	0
500,000 to 999,999	0
250,000 to 499,999	1,000
100,000 to 249,999	1,000
50,000 to 99,999	2,000
25,000 to 49,999	4,000
10,000 to 24,999	12,000
5,000 to 9,999	21,000
2,500 to 4,999	39,000
Under 2,500	151,000
Total	231,000
Percent of total firefighters	21%

The above projections are based on 8,243 departments reporting yes on Question 13a and reporting on Question 13b and assume "Most" = 2/3 and "Some" = 1/3. Numbers are estimated to the nearest 1,000 and may not add to totals due to rounding. See Tables 10 and 11.

Q. 13b: If [structural firefighting is a role your department performs] how many of your personnel who perform this duty have received formal training (not just on-the job)? All, Most, Some, None.

Table 12 (p. 35) indicates what levels of certification are held by some or all of the firefighters who perform structural firefighting, by department. Using the same methods used to construct Table C, but applying them only to departments that did not indicate any type of certification provided, 153,000 firefighters are estimated to serve in fire departments where no certification of firefighters as Firefighter Level I or II has taken place.

None of these firefighters were in fire departments protecting at least 50,000 population. Most of the firefighters in departments with no certification for structural firefighting were in rural fire departments and so were almost certainly volunteer firefighters.

Note that there may be other firefighters – possibly many other firefighters – who lack certification serving in departments where some firefighters are certified. These firefighters are not reflected in the 153,000 figure cited above. Conversely, some departments where no one is certified may be providing a local equivalent of certification.

The breakdown by community size is shown in Table D.

Table D. Estimated Number of Firefighters Involved in Structural Firefighting Serving in Fire Departments Where No One is Certified, by Size of Community Protected (Q. 13c)

Population Protected	Estimated Firefighters Lacking Certification
1,000,000 or more	0
500,000 to 999,999	0
250,000 to 499,999	0
100,000 to 249,999	0
50,000 to 99,999	0
25,000 to 49,999	3,000
10,000 to 24,999	7,000
5,000 to 9,999	11,000
2,500 to 4,999	21,000
Under 2,500	111,000
Total	153,000
Percent of total firefighters	14%

The above projections are based on 8,359 departments reporting yes on Question 13a and reporting on Question 13c. Numbers are estimated to the nearest 1,000 and may not add to totals due to rounding. See Tables 10 and 12.

Q. 13c: [If structural firefighting is a role your department performs,] have any of your personnel been certified to any of the following levels? Firefighter Level I and II.

Emergency Medical Service

Table 13 (p. 36) asks whether emergency medical service (EMS) is within the scope of the fire department. Roughly one-third (35%) of departments say no, mostly in smaller

21

communities. However, even for rural fire departments, protecting fewer than 2,500 population, the majority of fire departments now provide EMS.

Table 14 (p. 37) asks how many of the assigned personnel in departments responsible for EMS have received formal training. The breakdown by community size is given in Figure 2 and Table E, in terms of percent of personnel performing this duty who lack formal training. For communities of 25,000 population or more, fewer than 10% of personnel involved in providing EMS are estimated to lack formal training. For small communities of less than 5,000 population, roughly one-third of involved personnel are estimated to lack formal training.

These percentages cannot be safely converted to estimates of numbers of personnel lacking formal training. Table A provides statistics on numbers of firefighters by size of community and Table 13 provides statistics on fraction of departments providing EMS by size of community, but no table is available to indicate what fraction of total firefighters are assigned to EMS duties or how many non-firefighters are also assigned to those duties. Therefore, the percentages shown in Table E apply to a base of involved personnel whose size we do not know.

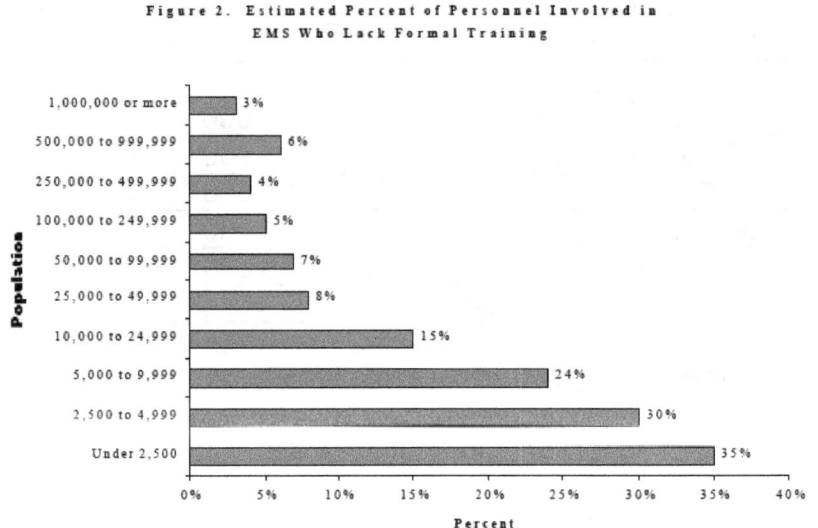

Figure 2. Estimated Percent of Personnel Involved in
EMS Who Lack Formal Training

Table 15 (p. 38) indicates certification of personnel who perform EMS. The question asked whether any personnel had been certified to any of several possible levels. The columns of Table 15 are defined by combinations of the four levels of certification.

Since the four levels are progressive, with each level incorporating the skills and knowledge of the previous level, it is reasonable to assume that a combination answer (e.g., First Responder and Basic Life Support) indicates that some personnel in the department are certified to one of the levels and other personnel are certified to another level. By contrast, a department that responds with only one level presumably has all its certified personnel certified to that one level. In every case, it is possible that some assigned personnel are not certified to any level.

Table E. Estimated Percentage of Personnel Involved in EMS Who Lack Formal Training, by Size of Community Protected (Q. 14b)

Population Protected	Estimated % of Personnel Lacking Formal Training
1,000,000 or more	3%
500,000 to 999,999	6%
250,000 to 499,999	4%
100,000 to 249,999	5%
50,000 to 99,999	7%
25,000 to 49,999	8%
10,000 to 24,999	15%
5,000 to 9,999	24%
2,500 to 4,999	30%
Under 2,500	35%
Total	27%

The above projections are based on 5,597 departments reporting yes on Question 14a and reporting on Question 14b and assume "Most" = 2/3 and "Some" = 1/3. See Tables 13 and 14.

Q. 14b: If [emergency medical services (EMS) is a role your department performs], how many of your personnel who perform this duty have received formal training (not just on-the job)? All, Most, Some, None.

Table 15 indicates that almost no departments performing EMS are completely lacking in certified personnel. Conversely, no departments reported that all their certified personnel were certified to the level of Paramedic, the highest level of certification, and almost no departments reported that all their certified personnel were certified to the level of Advanced Life Support, the second highest level of certification.

Hazardous Material Response

Table 16 (p. 39) asks whether hazardous material response is within the scope of the fire department. Nearly one-fourth (23%) of departments say no, and the ones saying no are mostly smaller communities. Even for rural fire departments, protecting fewer than

2,500 population, roughly two-thirds of fire departments now provide hazardous material response.

Table 17 (p. 40) asks how many of the assigned personnel in departments responsible for hazardous material response have received formal training. Requirements of the US Environmental Protection Agency (EPA) and the US Occupational Safety and Health Administration (OSHA) specify that all assigned personnel must have formal training. Table 17 indicates only 27% of departments are compliant with these requirements, including two-thirds or more of departments protecting communities of at least 50,000 population – where most departments are all- or mostly-career – and one-seventh of departments protecting rural communities.

The breakdown of lack of training by community size is given in Table F, in terms of percent of personnel performing this duty who lack training, by size of community protected.

Table F. Estimated Percentage of Personnel Involved in Hazardous Material Response Who Lack Formal Training by Size of Community Protected (Q. 15b)

Population Protected	Estimated % of Personnel Lacking Formal Training
1,000,000 or more	18%
500,000 to 999,999	9%
250,000 to 499,999	10%
100,000 to 249,999	15%
50,000 to 99,999	16%
25,000 to 49,999	20%
10,000 to 24,999	27%
5,000 to 9,999	35%
2,500 to 4,999	42%
Under 2,500	50%
Total	40%

The above projections are based on 3,832 departments reporting yes on Question 15a and reporting on Question 15b, and assume "Most" = 2/3 and "Some" = 1/3. See Tables 16 and 17.

Q. 15b: If [hazardous materials response is a role your department performs], how many of your personnel who perform this duty have received formal training (not just on-the-job)? All, Most, Some, None.

These percentages cannot be safely converted to estimates of numbers of personnel lacking training. Table A provides statistics on numbers of firefighters by size of community and Table 16 on fraction of departments providing hazardous material response by size of community, but no table is available to indicate what fraction of total firefighters are assigned to hazardous material response duties or how many non-firefighters may also be assigned to those duties. Therefore, the percentages shown in Table F apply to a base of involved personnel whose size we do not know.

Table 18 (p. 41) indicates certification of firefighters who perform hazardous material response. The columns of Table 18 are defined by combinations of the three levels of certification. Since the three levels are progressive, with each level incorporating the skills and knowledge of the previous level, it is reasonable to assume that a combination answer (e.g., Awareness and Technician) indicates that some personnel are certified to one level and other personnel are certified to another level. By contrast, a department that responds with only one level presumably has all its certified personnel certified to that level. In every case, it is possible that some assigned personnel are not certified to any level.

Except for rural communities, almost no departments performing hazardous material response are completely lacking in certified personnel (less than 4% of departments in each population interval, except for 7% for communities of less than 2,500 population). At the other end, almost no departments (3% of departments) have all involved personnel certified to the highest level, which is Technician, and less than one-fifth (18%) have all involved personnel certified to at least the second highest level, which is Operational.

Wildland Firefighting

Table 19 (p. 42) asks whether wildland firefighting is within the scope of the fire department. Roughly one-sixth (16%) of departments say no, with the percentage falling to 11% in smaller communities and over 30% for departments protecting communities of at least 25,000 population. Even for the most urban fire departments, at least three-fifths of fire departments provide wildland firefighting.

Table 20 (p. 43) asks how many of the assigned personnel in departments responsible for wildland firefighting have received formal training. The breakdown of lack of formal training by community size is given in Table G, in terms of percent of personnel performing this duty lacking training, by size of community protected.

Table G indicates roughly one-fourth to one-third of assigned personnel lack formal training, depending on community size. The percent of personnel lacking training is larger for smaller communities, which are also more likely to provide wildland firefighting as a service.

These percentages cannot be safely converted to estimates of numbers of personnel lacking training. Table A provides statistics on numbers of first-response personnel by size of community and Table 19 provides statistics on fraction of departments providing wildland firefighting by size of community, but no table is available to indicate what fraction of total first-response personnel are assigned to wildland firefighting duties. Most departments will use special teams for this service, which means that the number of personnel involved in performing this duty will typically be less than the total number of firefighting personnel. Therefore, the percentages shown in Table G apply to a base of involved personnel whose size we do not know.

Table G. Estimated Percentage of Personnel Involved in Wildland Firefighting Who Lack Formal Training by Size of Community Protected (Q. 16b)

Population Protected	Estimated % of Personnel Lacking Formal Training
1,000,000 or more	33%
500,000 to 999,999	18%
250,000 to 499,999	25%
100,000 to 249,999	30%
50,000 to 99,999	28%
25,000 to 49,999	33%
10,000 to 24,999	35%
5,000 to 9,999	37%
2,500 to 4,999	40%
Under 2,500	45%
Total	41%

The above projections are based on 6,680 departments reporting yes on Question 16a and reporting on Question 16b and assume "Most" = 2/3 and "Some" = 1/3. See Tables 19 and 20.

Q. 16b: If [wildland firefighting is a role your department performs], how many of your personnel who perform this duty have received formal training (not just on-the-job)? All, Most, Some, None.

Technical Rescue

Table 21 (p. 44) asks whether technical rescue is within the scope of the fire department. Nearly half (44%) of departments say no, mostly in smaller communities. However, even for rural fire departments, protecting fewer than 2,500 population, nearly half of fire departments now provide technical rescue.

Table 22 (p. 45) asks how many of the assigned personnel in departments responsible for technical rescue service have received formal training. The breakdown of lack of training by community size is given in Table H, in terms of percent of personnel performing this duty lacking training, by size of community protected.

For communities with at least 500,000 population, one-fourth to one-third of personnel performing this duty lacked formal training. For communities with 5,000 to 499,999 population, nearly half lacked formal training. For communities with less than 5,000 population, just over half lacked formal training.

Table H. Estimated Percentage of Personnel Involved in Technical Rescue Service Who Lack Formal Training by Size of Community Protected (Q. 17b)

Population Protected	Estimated % of Personnel Lacking Formal Training
1,000,000 or more	27%
500,000 to 999,999	25%
250,000 to 499,999	44%
100,000 to 249,999	40%
50,000 to 99,999	40%
25,000 to 49,999	44%
10,000 to 24,999	45%
5,000 to 9,999	46%
2,500 to 4,999	51%
Under 2,500	56%
Total	53%

The above projections are based on 5,072 departments reporting on Question 17b and assume "Most" = 2/3 and "Some" = 1/3. See Tables 21 and 22.

Q. 17b: If [technical rescue is a role your department performs], how many of your personnel who perform this duty have received formal training (not just on-the-job)? All, Most, Some, None.

These percentages cannot be safely converted to estimates of numbers of personnel lacking formal training.

Table A provides statistics on numbers of firefighters by size of community and Table 21 provides statistics on fraction of departments providing technical rescue by size of community, but no table is available to indicate what fraction of total firefighters are assigned to technical rescue or how many non-firefighters may be assigned to such duties.

Most departments will use special teams for this service, which means that the number of personnel involved in performing this duty will typically be less than the total number of firefighting personnel.

Therefore, the percentages shown in Table H apply to a base of involved personnel whose size we do not know.

Programs to Maintain and Protect Firefighter Health

Table 23 (p. 46) indicates whether departments have a program to maintain basic firefighter fitness and health, such as is required in NFPA 1500, *Standard on Fire Department Occupational Safety and Health Program.*

Only one-fifth of fire departments indicate that they have such a program, although half or more of communities with at least 50,000 population report programs.

Figure 3 estimates what percentage of firefighters, career or volunteer, are in departments without such programs.

In the largest communities, those with populations of 250,000 or more, 30-40% of firefighters are estimated to work in fire departments without programs to maintain basic firefighter fitness and health.

In the smallest communities, those with populations of less than 10,000, at least three-fourths of firefighters are estimated in serve in fire departments without such programs.

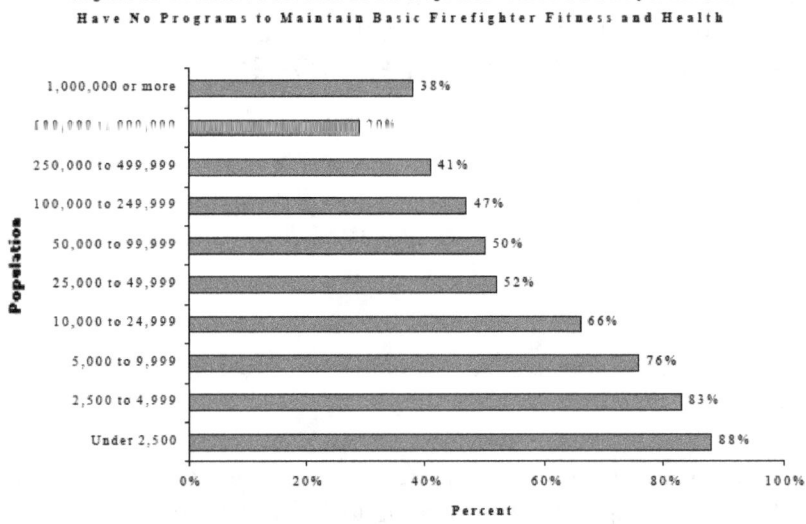

Figure 3. Estimated Percent of Firefighters Whose Fire Departments Have No Programs to Maintain Basic Firefighter Fitness and Health

Table I estimates how many firefighters, career or volunteer, are in departments without such programs. Because such a large share of total firefighters serve as volunteers in smaller communities, which are the same communities where most fire departments do not have programs to maintain basic firefighter fitness and health, the estimated total of 792,000 firefighters without such programs represents roughly three-fourths of the estimated total number of firefighters.

Table I. Estimated Number of Firefighters in Fire Departments With No Program to Maintain Basic Firefighter Fitness and Health by Size of Community Protected (Q. 18)

Population Protected	Estimated Firefighters Without Program to Maintain Fitness
1,000,000 or more	13,000
500,000 to 999,999	10,000
250,000 to 499,999	13,000
100,000 to 249,999	22,000
50,000 to 99,999	24,000
25,000 to 49,999	36,000
10,000 to 24,999	83,000
5,000 to 9,999	94,000
2,500 to 4,999	134,000
Under 2,500	363,000
Total	792,000
Percent of total firefighters	73%

The above projections are based on 8,267 departments reporting on Question 18. Numbers are shown to the nearest 1,000 and may not sum to totals due to rounding. See Table 23.

Q. 18: Does your department have a program to maintain basic firefighter fitness and health (e.g., as required in NFPA 1500)?

Table 24 (p. 47) indicates that nearly two-thirds of fire departments have programs for infectious disease control, including more than 95% of departments protecting communities with at least 50,000 population.

Table 7
For All- or Mostly-Volunteer Departments
Average Number of Volunteer Firefighters Who Respond to a Mid-Day House Fire
Percent of Departments by Community Size
(Q. 10)

Average Number of Volunteer Firefighters Responding

Population of Community	1-2	3-4	5-9	10-14	15-19	20 or More	Total
25,000 to 49,999	2.6%	9.4%	23.0%	28.3%	15.2%	21.5%	100.0%
10,000 to 24,999	3.7	11.7	32.8	26.5	13.0	12.4	100.0
5,000 to 9,999	2.9	11.4	39.5	26.1	12.7	7.3	100.0
2,500 to 4,999	3.1	12.2	45.9	25.6	9.7	3.6	100.0
Under 2,500	3.0	18.3	48.0	22.1	6.1	2.5	100.0

Source: FEMA US Fire Administration 2002
        Survey of the Needs of the US Fire Service

A mostly-volunteer department might respond with some career firefighters as well, but this question asked only about volunteers responding.

The above projections are based on 6,237 departments reporting on Question 10 and comprised of all- or mostly-volunteer firefighters. Numbers may not add to totals due to rounding.

Q. 10: Average number of call/volunteer personnel who respond to a mid-day house fire (blank for actual number).

Table 8
For All- or Mostly-Career Departments
Number of Career Firefighters Assigned to an Engine/Pumper Apparatus
Percent of Departments by Community Size
(Q. 11)

Number of Career Firefighters Assigned to Engine/Pumper

Population of Community	1-2	3	4	5 or More	Total
1,000,000 or more	0.0%	0.0%	87.5%	12.5%	100.0%
500,000 to 000,009	0.0	40.0	56.7	3.3	100.0
250,000 to 499,999	2.3	38.6	54.5	4.5	100.0
100,000 to 249,999	5.5	50.6	39.6	4.2	100.0
50,000 to 99,999	10.0	66.0	21.4	2.6	100.0
25,000 to 49,999	17.9	63.7	18.2	1.8	100.0
10,000 to 24,999	30.5	42.2	23.5	2.4	100.0

Source: FEMA US Fire Administration 2002
Survey of the Needs of the US Fire Service

The above projections are based on 1,399 departments reporting on Question 11 and comprised of all- or mostly-career firefighters. Numbers may not add to totals due to rounding.

Q. 11: Number of on-duty career/paid personnel assigned to an engine/pumper (answers given as ranges shown).

Table 9
For All- or Mostly-Career Departments
Number of Career Firefighters Assigned to a Ladder/Aerial Apparatus
Percent of Departments by Community Size
(Q. 12)

Population of Community	Number of Career Firefighters Assigned to a Ladder/Aerial					
	1-2	3	4	5 or More	Not Applicable	Total
1,000,000 or more	0.0%	0.0%	62.5%	37.5%	0.0%	100.0%
500,000 to 999,999	0.0	33.3	50.0	16.7	0.0	100.0
250,000 to 499,999	4.5	29.5	61.4	4.5	0.0	100.0
100,000 to 249,999	15.2	34.5	43.0	3.6	3.6	100.0
50,000 to 99,999	22.2	44.4	20.6	2.9	9.8	100.0
25,000 to 49,999	40.4	28.6	12.9	1.6	16.5	100.0
10,000 to 24,999	49.4	15.4	7.8	0.2	27.1	100.0

Source: FEMA US Fire Administration 2002
Survey of the Needs of the US Fire Service

The above projections are based on 1,381 departments reporting on Question 12 and comprised of all- or mostly-career firefighters. Numbers may not add to totals due to rounding.

Q. 12: Number of on-duty career/paid personnel assigned to a ladder/aerial (answers given as ranges shown).

Table 10
Does Department Provide Structural Firefighting?
by Community Size
(Q. 13a)

| Population | Yes | | No | | Total | |
of Community	Number Depts	Percent	Number Depts	Percent	Number Depts	Percent
1,000,000 or more	13	100.0%	0	0.0%	13	100.0%
500,000 to 999,999	38	100.0	0	0.0	38	100.0
250,000 to 499,999	64	100.0	0	0.0	64	100.0
100,000 to 249,999	215	100.0	0	0.0	215	100.0
50,000 to 99,999	487	100.0	0	0.0	487	100.0
25,000 to 49,999	1,053	100.0	0	0.0	1,053	100.0
10,000 to 24,999	2,843	100.0	0	0.0	2,843	100.0
5,000 to 9,999	3,625	99.9	4	0.1	3,629	100.0
2,500 to 4,999	4,567	99.9	5	0.1	4,572	100.0
Under 2,500	13,244	98.5	196	1.5	13,440	100.0
Total	26,150	99.2	204	0.8	26,354	100.0

Source: FEMA US Fire Administration 2002
         Survey of the Needs of the US Fire Service

The above projections are based on 8,403 departments reporting on Question 13a. Numbers may not add to totals due to rounding.

Q. 13a: Is [structural firefighting] a role your department performs?

Table 11
For Departments That Provide Structural Firefighting
How Many Personnel Who Perform This Duty Have Received Formal Training?
by Community Size
(Q. 13b)

Population of Community	All		Most		Some		None		Total	
	Number Depts	Percent	Number Depts	Percent	Number Depts	Percent	Number Depts	Percent	Number Depts	Percent
1,000,000 or more	13	100.0%	0	0.0%	0	0.0%	0	0.0%	13	100.0%
500,000 to 999,999	38	100.0	0	0.0	0	0.0	0	0.0	38	100.0
250,000 to 499,999	59	93.0	5	7.0	0	0.0	0	0.0	64	100.0
100,000 to 249,999	204	94.9	11	5.1	0	0.0	0	0.0	215	100.0
50,000 to 99,999	440	90.3	47	9.7	0	0.0	0	0.0	487	100.0
25,000 to 49,999	893	84.8	131	12.4	29	2.8	0	0.0	1,053	100.0
10,000 to 24,999	2,146	75.5	558	19.6	137	4.8	2	0.1	2,843	100.0
5,000 to 9,999	2,163	59.7	1,107	30.6	334	9.2	21	0.6	3,625	100.0
2,500 to 4,999	2,105	46.1	1,706	37.4	716	15.7	38	0.8	4,565	100.0
Under 2,500	3,557	26.9	5,048	38.1	4,183	31.6	453	3.4	13,241	100.0
Total	11,616	44.4	8,601	32.9	5,417	20.7	513	2.0	26,150	100.0

Source: FEMA US Fire Administration 2002
Survey of the Needs of the US Fire Service

The above projections are based on 8,243 departments reporting yes to Question 13a and also reporting on this question. Numbers may not add to totals due to rounding.

Q. 13b: If [structural firefighting is a role your department performs], how many of your personnel who perform this duty have received formal training (not just on-the-job)?

34

Table 12
For Departments That Provide Structural Firefighting,
Level That Personnel Who Perform This Duty Have Been Certified to
Percent of Departments by Community Size
(Q. 13c)

Population of Community	No Certification	Firefighter Level 1	Firefighter Level 2	Both Levels	Total Departments
1,000,000 or more	0.0%	11.1%	55.5%	33.3%	100.0%
500,000 to 999,999	0.0	6.7	16.7	76.7	100.0
250,000 to 499,999	0.0	19.1	35.8	45.1	100.0
100,000 to 249,999	0.0	12.1	28.7	59.2	100.0
50,000 to 99,999	0.0	9.7	33.3	57.0	100.0
25,000 to 49,999	4.0	8.5	36.3	51.1	100.0
10,000 to 24,999	5.4	13.4	26.3	54.8	100.0
5,000 to 9,999	8.9	22.1	18.4	50.7	100.0
2,500 to 4,999	13.2	30.4	13.3	43.1	100.0
Under 2,500	26.8	38.3	8.5	26.4	100.0
Total	14.5	29.0	15.0	41.0	100.0

Source: FEMA US Fire Administration 2002
        Survey of the Needs of the US Fire Service

The above projections are based on 8,359 departments reporting yes to Question 13a
and also reporting on Question 13c. Numbers may not add to totals due to rounding.

Q. 13c: Have any of your personnel been certified to any of the following levels? Firefighter
Level I, II

Table 13
Does Department Provide Emergency Medical Service (EMS)?
by Community Size
(Q. 14a)

	Yes		No		Total	
Population of Community	Number Depts	Percent	Number Depts	Percent	Number Depts	Percent
1,000,000 or more	13	100.0%	0	0.0%	13	100.0%
500,000 to 999,999	38	100.0	0	0.0	38	100.0
250,000 to 499,999	63	98.4	1	1.6	64	100.0
100,000 to 249,999	209	97.2	6	2.9	215	100.0
50,000 to 99,999	452	92.8	35	7.2	487	100.0
25,000 to 49,999	931	88.4	122	11.6	1,053	100.0
10,000 to 24,999	2,164	76.1	679	23.9	2,843	100.0
5,000 to 9,999	2,491	68.7	1,137	31.3	3,629	100.0
2,500 to 4,999	2,976	65.1	1,596	34.9	4,572	100.0
Under 2,500	7,725	57.5	5,716	42.5	13,440	100.0
Total	17,058	64.7	9,296	35.3	26,354	100.0

Source: FEMA US Fire Administration 2002
Survey of the Needs of the US Fire Service

The above projections are based on 8,306 departments reporting on Question 14a. Numbers may not add to totals due to rounding.

Q. 14a: Is [emergency medical service] a role your department performs?

Table 14
For Departments That Provide Emergency Medical Service
How Many Personnel Who Perform This Duty Have Received Formal Training?
by Community Size
(Q. 14b)

Population of Community	All		Most		Some		None		Total	
	Number Depts	Percent	Number Depts	Percent	Number Depts	Percent	Number Depts	Percent	Number Depts	Percent
1,000,000 or more	12	90.0%	1	10.0%	0	0.0%	0	0.0%	13	100.0%
500,000 to 999,999	31	81.6	7	18.4	0	0.0	0	0.0	38	100.0
250,000 to 499,999	56	88.9	6	9.5	1	1.6	0	0.0	63	100.0
100,000 to 249,999	182	87.1	22	10.6	5	2.4	0	0.0	209	100.0
50,000 to 99,999	372	82.4	66	14.6	14	3.0	0	0.0	452	100.0
25,000 to 49,999	757	81.3	125	13.4	49	5.2	0	0.0	931	100.0
10,000 to 24,999	1,398	64.9	538	25.0	225	10.5	2	0.1	2,164	100.0
5,000 to 9,999	1,211	48.6	785	31.5	490	19.7	3	0.1	2,491	100.0
2,500 to 4,999	1,194	40.1	928	31.2	840	28.2	15	0.5	2,976	100.0
Under 2,500	2,697	34.9	1,870	24.2	3,115	40.3	42	0.5	7,725	100.0
Total	7,905	46.3	4,344	25.5	4,737	27.8	61	0.4	17,058	100.0

Source: FEMA US Fire Administration 2002
Survey of the Needs of the US Fire Service

The above projections are based on 5,597 departments reporting yes to Question 14a and also reporting on this question. Numbers may not add to totals due to rounding.

Q. 14b: If [emergency medical service is a role your department performs], how many of your personnel who perform this duty have received formal training (not just on-the-job)?

Table 15
For Departments That Provide Emergency Medical Service
Level That Personnel Have Been Certified to
For Departments by Community Size (Percent)
(Q.14c)

Population of Community	None	First Responder	Basic Life Support	First Responder Basic Life Support	First Responder Basic Life Support Advanced Life Support	Basic Life Support Advanced Life Support Paramedic	First Responder Advanced Life Support Paramedic	Advanced Life Support Paramedic	Total
1,000,000 or more	0.0%	0.0%	11.1%	0.0%	44.4%	33.3%	11.1%	0.0%	100.0%
500,000 to 999,999	0.0	9.1	3.0	3.0	30.3	51.5	0.0	3.0	100.0
250,000 to 499,999	0.0	6.4	8.5	10.6	23.4	36.2	2.1	12.8	100.0
100,000 to 249,999	1.2	4.8	6.6	9.6	32.5	37.3	0.6	7.2	100.0
50,000 to 99,999	0.6	4.2	8.6	10.7	33.6	31.8	2.7	7.7	100.0
25,000 to 49,999	0.2	4.4	8.9	12.4	33.7	32.1	1.0	7.4	100.0
10,000 to 24,999	1.0	5.3	9.5	18.5	28.4	30.4	1.1	5.8	100.0
5,000 to 9,999	0.9	8.3	8.7	28.4	15.3	32.2	2.3	3.9	100.0
2,500 to 4,999	0.7	10.0	9.6	30.5	14.0	30.7	2.5	2.0	100.0
Under 2,500	1.3	18.4	14.5	35.0	6.9	19.2	3.1	1.4	100.0
Total	1.0	12.4	11.5	28.7	14.7	25.9	2.5	3.0	100.0

Source: FEMA U.S. Fire Administration 2002
Survey of the Needs of the U.S. Fire Service

The above projections are based on 4,952 departments reporting yes to Question 14a, and also reporting on this question. Numbers may not add to totals due to rounding.

Q. 14c: If [emergency medical service is a role your department performs], have any of your personnel been certified to any of the following levels?

38

Table 16
Does Department Provide Hazardous Material Response?
by Community Size
(Q. 15a)

| Population | Yes | | No | | Total | |
of Community	Number Depts	Percent	Number Depts	Percent	Number Depts	Percent
1,000,000 or more	13	100.0%	0	0.0%	13	100.0%
500,000 to 999,999	38	100.0	0	0.0	38	100.0
250,000 to 499,999	64	100.0	0	0.0	64	100.0
100,000 to 249,999	206	95.8	9	4.2	215	100.0
50,000 to 99,999	466	95.7	21	4.3	487	100.0
25,000 to 49,999	999	94.9	54	5.1	1,053	100.0
10,000 to 24,999	2,611	91.8	232	8.2	2,843	100.0
5,000 to 9,999	3,176	87.5	453	12.5	3,629	100.0
2,500 to 4,999	3,760	82.2	812	17.8	4,572	100.0
Under 2,500	9,025	67.2	4,414	32.8	13,440	100.0
Total	20,360	77.3	5,994	22.7	26,354	100.0

Source: FEMA US Fire Administration 2002
Survey of the Needs of the US Fire Service

The above table projections are based on 8,361 departments reporting on Question 15a.
Numbers may not add to totals due to rounding.

Q. 15a: Is [hazardous materials response] a role your department performs?

Table 17
For Departments That Provide Hazardous Material Response
How Many Personnel Who Perform This Duty Have Received Formal Training?
by Community Size
(Q. 15b)

Population of Community	All		Most		Some		None		Total	
	Number Depts	Percent	Number Depts	Percent	Number Depts	Percent	Number Depts	Percent	Number Depts	Percent
1,000,000 or more	9	69.2%	1	7.7%	3	23.1%	0	0.0%	13	100.0%
500,000 to 999,999	31	81.6	4	10.5	3	7.9	0	0.0	38	100.0
250,000 to 499,999	52	81.3	5	7.8	7	10.9	0	0.0	64	100.0
100,000 to 249,999	146	70.8	26	12.6	35	16.8	0	0.0	206	100.0
50,000 to 99,999	320	68.6	70	15.0	75	16.1	1	0.3	466	100.0
25,000 to 49,999	594	59.5	203	20.3	199	20.0	2	0.2	999	100.0
10,000 to 24,999	1,178	45.0	737	28.3	682	26.2	14	0.5	2,611	100.0
5,000 to 9,999	1,015	32.0	1,066	33.6	1,066	33.6	29	0.9	3,176	100.0
2,500 to 4,999	811	21.6	1,210	32.2	1,670	44.4	68	1.8	3,760	100.0
Under 2,500	1,296	14.4	2,371	26.3	5,016	55.6	338	3.8	9,025	100.0
Total	5,449	26.8	5,690	28.0	8,760	43.0	452	2.2	20,360	100.0

Source: FEMA US Fire Administration 2002
Survey of the Needs of the US Fire Service

The above projections are based on 3,832 departments reporting yes to Questions 15a and also reporting on this question. Numbers may not add to totals due to rounding.

Q. 15b: If [hazardous materials response is a role your department performs], how many of your personnel who perform this duty have received formal training (not just on-the-job)?

Table 18
For Departments That Provide Hazardous Material Response
Level That Personnel Who Perform This Duty Have Been Certified to
Percent of Departments by Community Size

(Q. 15c)

Population of Community	None	Awareness	Operational	Technician	Awareness Operational	Awareness Technician	Operational Technician	Awareness Operational Technician	Total
1,000,000 or more	0.0%	0.0%	0.0%	11.1%	11.1%	0.0%	11.1%	66.6%	100.0%
500,000 to 999,999	3.0	0.0	0.0	12.1	3.0	6.1	15.2	60.6	100.0
250,000 to 499,999	2.2	4.4	2.2	8.9	6.7	0.0	22.2	53.3	100.0
100,000 to 249,999	3.0	4.2	4.2	7.8	9.0	0.6	21.7	49.4	100.0
50,000 to 99,999	1.2	5.7	8.4	6.3	12.3	2.1	16.8	47.3	100.0
25,000 to 49,999	1.0	7.8	12.4	4.8	13.9	2.1	15.5	42.6	100.0
10,000 to 24,999	1.2	11.3	12.2	5.4	23.4	2.1	10.4	33.9	100.0
5,000 to 9,999	3.0	19.2	11.6	2.9	30.7	1.6	4.4	26.6	100.0
2,500 to 4,999	3.2	28.2	9.5	2.1	33.4	1.8	2.9	19.0	100.0
Under 2,500	7.2	43.7	10.0	1.7	24.7	1.6	1.2	10.0	100.0
Total	4.5	29.5	10.4	2.8	26.0	1.7	4.5	20.4	100.0

Source: FEMA US Fire Administration 2002
Survey of the Needs of the US Fire Service

The above projections are based on 4,945 departments reporting yes to Question 15a and also reporting on this question. Numbers may not add to totals due to rounding.

Q. 15c: If [hazardous material response is a role your department performs], have any of your personnel been certified to any of the following levels?

41

Table 19
Does Department Provide Wildland Firefighting?
by Community Size
(Q. 16a)

	Yes		No		Total	
Population of Community	Number Depts	Percent	Number Depts	Percent	Number Depts	Percent
1,000,000 or more	9	69.2%	4	30.8%	13	100.0%
500,000 to 999,999	26	68.4	12	31.6	38	100.0
250,000 to 499,999	42	65.6	22	34.4	64	100.0
100,000 to 249,999	148	68.8	67	31.2	215	100.0
50,000 to 99,999	323	66.4	164	33.6	487	100.0
25,000 to 49,999	659	62.6	394	37.4	1,053	100.0
10,000 to 24,999	2,026	71.3	817	28.7	2,843	100.0
5,000 to 9,999	3,013	83.0	616	17.0	3,629	100.0
2,500 to 4,999	3,939	86.2	633	13.8	4,572	100.0
Under 2,500	11,938	88.8	1,502	11.2	13,440	100.0
Total	22,126	84.0	4,228	16.0	26,354	100.0

Source: FEMA US Fire Administration 2002
    Survey of the Needs of the US Fire Service

The above projections are based on 8,348 departments reporting on Question 16a. Numbers may not add to totals due to rounding.

Q. 16a: Is [wildland firefighting] a role your department performs?

Table 20
For Departments That Provide Wildland Firefighting
How Many Personnel Who Perform This Duty Have Received Formal Training?
by Community Size

(Q. 16b)

Population of Community	All		Most		Some		None		Total	
	Number Depts	Percent	Number Depts	Percent	Number Depts	Percent	Number Depts	Percent	Number Depts	Percent
1,000,000 or more	4	44.4%	1	11.1%	4	44.4%	0	0.0%	9	100.0%
500,000 to 999,999	20	76.9	0	0.0	4	15.4	2	7.7	26	100.0
250,000 to 499,999	24	57.1	4	9.5	14	33.3	0	0.0	42	100.0
100,000 to 249,999	72	48.6	29	19.6	38	25.7	9	6.1	148	100.0
50,000 to 99,999	153	47.4	82	25.4	74	22.9	14	4.2	323	100.0
25,000 to 49,999	283	42.9	145	22.0	178	27.0	51	7.8	659	100.0
10,000 to 24,999	731	36.1	596	29.4	575	28.4	124	6.2	2,026	100.0
5,000 to 9,999	916	30.4	1,029	34.1	882	29.3	186	6.2	3,013	100.0
2,500 to 4,999	965	24.5	1,460	37.1	1,240	31.5	270	6.8	3,939	100.0
Under 2,500	2,333	19.5	4,300	36.0	4,088	34.2	1,217	10.2	11,938	100.0
Total	5,499	24.9	7,652	34.6	7,095	32.1	1,873	8.5	22,126	100.0

Source: FEMA US Fire Administration 2002
Survey of the Needs of the US Fire Service

The above projections are based on 6,680 departments reporting yes to Question 16a and also reporting on this question. Numbers may not add to totals due to rounding.

Q. 16b: If [wildland firefighting is a role your department performs], how many of your personnel who perform this duty have received formal training (not just on-the-job)?

43

Table 21
Does Department Provide Technical Rescue Service?
by Community Size
(Q. 17a)

Population	Yes		No		Total	
	Number		**Number**		**Number**	
of Community	**Depts**	**Percent**	**Depts**	**Percent**	**Depts**	**Percent**
1,000,000 or more	13	100.0%	0	0.0%	13	100.0%
500,000 to 999,999	38	100.0	0	0.0	38	100.0
250,000 to 499,999	64	100.0	0	0.0	64	100.0
100,000 to 249,999	200	93.0	15	7.0	215	100.0
50,000 to 99,999	423	86.9	64	13.1	487	100.0
25,000 to 49,999	877	83.3	176	16.7	1,053	100.0
10,000 to 24,999	2,065	72.6	777	27.4	2,843	100.0
5,000 to 9,999	2,432	67.0	1,197	33.0	3,629	100.0
2,500 to 4,999	2,674	58.5	1,897	41.5	4,572	100.0
Under 2,500	5,853	43.5	7,587	56.5	13,440	100.0
Total	14,642	55.6	11,712	44.4	26,354	100.0

Source: FEMA US Fire Administration 2002
    Survey of the Needs of the US Fire Service

The above projections are based on 8,258 departments reporting on Question 17a. Numbers may not add to totals due to rounding.

Q. 17a: Is [technical rescue] a role your department performs?

Table 22
For Departments That Provide Technical Rescue Service
How Many Personnel Who Perform This Duty Have Received Formal Training?
by Community Size

(Q. 17b)

Population of Community	All		Most		Some		None		Total	
	Number Depts	Percent	Number Depts	Percent	Number Depts	Percent	Number Depts	Percent	Number Depts	Percent
1,000,000 or more	7	53.8%	0	0.0%	6	41.2%	0	0.0%	13	100.0%
500,000 to 999,999	20	52.6	7	18.4	11	28.9	0	0.0	38	100.0
250,000 to 499,999	18	28.1	7	10.9	39	60.9	0	0.0	64	100.0
100,000 to 249,999	58	29.0	47	23.5	94	47.0	1	0.5	200	100.0
50,000 to 99,999	109	25.8	113	26.7	200	47.3	0	0.0	423	100.0
25,000 to 49,999	198	22.6	214	24.4	460	52.6	3	0.4	877	100.0
10,000 to 24,999	394	19.1	616	29.9	1,019	49.3	36	1.8	2,065	100.0
5,000 to 9,999	345	14.2	819	33.8	1,223	50.4	40	1.6	2,432	100.0
2,500 to 4,999	261	9.8	827	30.9	1,470	55.1	112	4.2	2,674	100.0
Under 2,500	395	6.7	1,510	25.8	3,482	59.5	464	8.0	5,853	100.0
Total	1,804	12.3	4,157	28.4	8,012	57.8	656	4.5	14,642	100.0

Source: FEMA US Fire Administration 2002
Survey of the Needs of the US Fire Service

The above projections are based on 5,072 departments reporting yes to Question 17a and also reporting on this question. Numbers may not add to totals due to rounding.

Q. 17b: If [technical rescue is a role your department performs], how many of your personnel who perform this duty have received formal training (not just on-the-job)?

45

Table 23
Does Department Have a Program
to Maintain Basic Firefighter Fitness and Health?
by Community Size
(Q. 18)

	Yes		No		Total	
Population of Community	Number Depts	Percent	Number Depts	Percent	Number Depts	Percent
1,000,000 or more	8	61.5%	5	38.5%	13	100.0%
500,000 to 999,999	27	71.1	11	28.9	38	100.0
250,000 to 499,999	38	59.4	26	40.6	64	100.0
100,000 to 249,999	115	53.5	100	46.5	215	100.0
50,000 to 99,999	243	49.9	244	50.1	487	100.0
25,000 to 49,999	501	47.6	552	52.4	1,053	100.0
10,000 to 24,999	962	33.8	1,881	66.2	2,843	100.0
5,000 to 9,999	873	24.1	2,756	75.9	3,629	100.0
2,500 to 4,999	798	17.5	3,774	82.5	4,572	100.0
Under 2,500	1,639	12.2	11,801	87.8	13,440	100.0
Total	5,205	19.8	21,149	80.2	26,354	100.0

Source: FEMA US Fire Administration 2002
Survey of the Needs of the US Fire Service

The above projections are based on 8,267 departments reporting on Question 18. Numbers may not add to totals due to rounding.

Q. 18: Does your department have a program to maintain basic firefighter fitness and health (e.g., as required in NFPA 1500)?

Table 24
Does Department Have a
Program for Infectious Disease Control?
by Community Size
(Q. 19)

Population	Yes		No		Total	
	Number		Number		Number	
of Community	Depts	Percent	Depts	Percent	Depts	Percent
1,000,000 or more	13	100.0%	0	0.0%	13	100.0%
500,000 to 999,999	38	100.0	0	0.0	38	100.0
250,000 to 499,999	63	98.4	1	1.6	64	100.0
100,000 to 249,999	211	98.1	4	1.9	215	100.0
50,000 to 99,999	465	95.5	22	4.5	487	100.0
25,000 to 49,999	989	93.9	64	6.1	1,053	100.0
10,000 to 24,999	2,492	87.6	351	12.4	2,843	100.0
5,000 to 9,999	2,917	80.4	712	19.6	3,629	100.0
2,500 to 4,999	3,169	69.3	1,403	30.7	4,572	100.0
Under 2,500	6,616	49.2	6,823	50.8	13,440	100.0
Total	16,973	64.4	9,381	35.6	26,354	100.0

Source: FEMA US Fire Administration 2002
    Survey of the Needs of the US Fire Service

The above projections are based on 8,348 departments reporting on Question 19. Numbers may not add to totals due to rounding.

Q. 19: Does your department have a program for infectious disease control?

FIRE PREVENTION AND CODE ENFORCEMENT

Some of the greatest value delivered by the US fire services comes in activities that prevent fires and other emergencies from occurring or that moderate their severity when they do occur.

Questions 20-22 provide information on a number of such programs, all of which were recognized as candidates for Federal assistance in PL 106-398.

Table 25 (p. 54) indicates what percentage of fire departments, by community size, reported having each of six specific fire prevention or code enforcement programs.

Table J indicates the number of fire departments lacking these programs and estimates the number of people living in communities protected by fire departments that do not conduct such programs.

Table J. Number of Fire Departments and Estimated Total Population Protected by Those Fire Departments Where Selected Fire Prevention or Code Enforcement Programs Are NOT Provided, by Size of Community Protected (Q. 20)

1. Plans Review

Population Protected	Number of Departments Without Program	Population Protected by Departments Without Program
1,000,000 or more	4	8,400,000
500,000 to 999,999	5	2,800,000
250,000 to 499,999	4	1,300,000
100,000 to 249,999	23	3,200,000
50,000 to 99,999	38	2,400,000
25,000 to 49,999	157	5,300,000
10,000 to 24,999	1,467	23,100,000
5,000 to 9,999	976	7,000,000
2,500 to 4,999	3,027	13,400,000
Under 2,500	10,671	17,100,000
Total	16,372	83,900,000
Percent of US total	62%	29%

The above projections are based on 7,159 departments reporting on Question 20. Population estimates are shown to the nearest 100,000 and may not add to totals due to rounding. See Table 25.

2. Permit Approval

Population Protected	Number of Departments Without Program	Population Protected by Departments Without Program
1,000,000 or more	5	11,200,000
500,000 to 999,999	3	2,100,000
250,000 to 499,999	12	3,800,000
100,000 to 249,999	34	4,800,000
50,000 to 99,999	126	7,700,000
25,000 to 49,999	392	13,200,000
10,000 to 24,999	2,181	34,400,000
5,000 to 9,999	2,054	14,700,000
2,500 to 4,999	3,973	17,600,000
Under 2,500	12,042	19,300,000
Total	20,822	128,800,000
Percent of US total	79%	45%

The above projections are based on 7,159 departments reporting on Question 20. Population estimates are shown to the nearest 100,000 and may not add to totals due to rounding. See Table 25.

3. Routine Testing of Active Systems (e.g., sprinkler, detection/alarm, smoke control)

Population Protected	Number of Departments Without Program	Population Protected by Departments Without Program
1,000,000 or more	7	14,100,000
500,000 to 999,999	6	3,500,000
250,000 to 499,999	20	6,300,000
100,000 to 249,999	58	8,100,000
50,000 to 99,999	160	9,800,000
25,000 to 49,999	403	13,600,000
10,000 to 24,999	2,161	34,100,000
5,000 to 9,999	2,098	15,000,000
2,500 to 4,999	3,822	16,900,000
Under 2,500	11,921	19,100,000
Total	20,656	140,500,000
Percent of US total	78%	49%

The above projections are based on 7,159 departments reporting on Question 20. Population estimates are shown to the nearest 100,000 and may not add to totals due to rounding. See Table 25.

4. Free Distribution of Home Smoke Alarms

Population Protected	Number of Departments Without Program	Population Protected by Departments Without Program
1,000,000 or more	4	8,400,000
500,000 to 999,999	7	4,200,000
250,000 to 499,999	12	3,800,000
100,000 to 249,999	56	7,800,000
50,000 to 99,999	148	9,100,000
25,000 to 49,999	400	13,500,000
10,000 to 24,999	1,820	28,700,000
5,000 to 9,999	1,847	13,200,000
2,500 to 4,999	3,214	14,200,000
Under 2,500	10,765	17,300,000
Total	18,273	120,100,000
Percent of US total	69%	42%

The above projections are based on 7,159 departments reporting on Question 20. Population estimates are shown to the nearest 100,000 and may not add to totals due to rounding. See Table 25.

5. Juvenile Firesetter Program

Population Protected	Number of Departments Without Program	Population Protected by Departments Without Program
1,000,000 or more	3	5,600,000
500,000 to 999,999	2	1,400,000
250,000 to 499,999	16	5,100,000
100,000 to 249,999	70	9,800,000
50,000 to 99,999	164	10,100,000
25,000 to 49,999	433	14,600,000
10,000 to 24,999	2,274	35,800,000
5,000 to 9,999	2,185	15,600,000
2,500 to 4,999	4,069	18,000,000
Under 2,500	12,660	20,300,000
Total	21,877	136,300,000
Percent of US total	83%	48%

The above projections are based on 7,159 departments reporting on Question 20. Population estimates are shown to the nearest 100,000 and may not add to totals due to rounding. See Table 25.

6. School Fire Safety Education Program Based on a National Model Curriculum

Population Protected	Number of Departments Without Program	Population Protected by Departments Without Program
1,000,000 or more	3	5,600,000
500,000 to 999,999	2	1,400,000
250,000 to 499,999	9	3,000,000
100,000 to 249,999	54	7,500,000
50,000 to 99,999	113	6,900,000
25,000 to 49,999	270	9,100,000
10,000 to 24,999	1,009	15,900,000
5,000 to 9,999	994	7,100,000
2,500 to 4,999	1,966	8,700,000
Under 2,500	7,983	12,800,000
Total	12,403	78,000,000
Percent of US total	47%	27%

The above projections are based on 7,159 departments reporting on Question 20. Population estimates are shown to the nearest 100,000 and may not add to totals due to rounding. See Table 25.

Q. 20: Which of the following programs or activities does your department conduct?

The program with the highest reported participation was school fire safety education programs based on a national model curriculum, where the majority of US fire departments reported conducting such a program. This is one of the few programs in this section where there is some independent information regarding participation, and that information would suggest that implementation of a school-based fire safety curriculum following a national model exists is closer to 5% of fire departments rather than the reported 53%.

This large discrepancy may be a matter of interpretation. For example, many fire departments provide presentations to schools (e.g., puppet shows) in which the content is based on the content of some national model fire safety curriculum. Such presentations would qualify as a program of the sort asked about, but they would in practice have little educational value. Therefore, considerable caution should be shown when considering the reported practices for this particular program.

Table 26 (p. 55) indicates which of several groups conduct fire-code inspections in the community. For communities of 50,000 population or more, at least 80% report the use of full-time fire department inspectors. The percentage drops to 70% for communities of 25,000 to 49,999 population, to 46% for communities of 10,000 to 24,999 population, to

20% for communities of 5,000 to 9,999 population, to 8% for communities of 2,500 to 4,999 population, and to 3% for communities with less than 2,500 population.

The next most commonly cited resource for conducting fire-code inspections was firefighters in-service. Only 30% of departments protecting communities of 1 million or more population cited the use of in-service firefighters. Then, 64% of communities of 500,000 to 999,999 population cited their use, falling to 56% for communities of 250,000 to 499,999, to 47-50% for communities of 25,000 to 249,999, to 36% for communities of 10,000 to 24,999, to 18% for communities of 5,000 to 9,999, and to 11-15% for communities with less than 5,000 population.

Building department inspectors were cited by 12-24% of departments by community size, and separate inspection departments were cited by 4-13%. "Other" inspectors – such as personnel from the state fire marshal's office – were cited mostly by smaller communities and were the principal inspection resource cited by communities with less than 10,000 population.

Of greatest concern were those departments that reported no one conducted fire-code inspections in their community. Roughly 7,200 fire departments reported this situation, nearly all of them departments serving rural communities (less than 2,500 population). These 7,200 departments protect an estimated 20,900,000 people, with two-fifths of that population located in rural communities.

Table 27 (p. 56) indicates which of several parties determines that a fire was deliberately set. Multiple answers were permitted. For communities of 50,000 population or more, fire department arson investigators were cited by at least 90% of departments in each population interval, and no one else was cited by more than 42% of departments.

In communities of 25,000 to 49,999 population, 80% of departments cited fire department arson investigators, 48% cited state arson investigators, 35% cited incident commanders, 26% cited the police department, and 22% cited regional arson task force investigators. These results indicate that multiple agency involvement is commonplace for these communities.

Communities of 10,000 to 24,999 population were the only ones in which two different agencies were each cited by a majority of departments – fire department arson investigators (64%) and state arson investigators (61%). Also, incident commanders were cited by 36% of departments and police departments by 25% of departments.

For communities of less than 10,000 population, state arson investigators were cited by at least 71% of departments in each population interval and were by far the principal resource for determination of intentional firesetting in those communities. Incident commanders were still frequently cited in those communities as well. Fire department arson investigators were cited by 40% of departments in communities of 5,000 to 9,999 population, by 29% of departments in communities of 2,500 to 4,999 population, and by 15% of departments in communities of less than 2,500 population.

Table 25
Which Programs or Activities Does Department Conduct?
by Community Size
(Q. 20)

Population of Community	Plans Review	Permit Approval	Routine Testing of Active Systems	Free Distribution of Smoke Alarms	Juvenile Firesetter Program	School Fire Safety Education Program	Other Prevention Program
1,000,000 or more	70.0%	60.0%	50.0%	70.0%	80.0%	80.0%	30.0%
500,000 to 999,999	87.8	90.9	84.8	81.8	93.9	93.9	45.4
250,000 to 499,999	93.8	81.3	68.8	81.3	75.0	85.4	37.5
100,000 to 249,999	89.5	84.0	72.9	74.0	67.4	75.1	34.3
50,000 to 99,999	92.1	74.2	67.1	69.7	66.3	76.8	28.2
25,000 to 49,999	85.1	62.8	61.7	62.0	58.9	74.4	28.7
10,000 to 24,999	48.4	23.3	24.0	36.0	20.0	64.5	22.5
5,000 to 9,999	73.1	43.4	42.2	49.1	39.8	72.6	25.2
2,500 to 4,999	33.8	13.1	16.4	29.7	11.0	57.0	16.8
Under 2,500	20.6	10.4	11.3	19.9	5.8	40.6	14.1
Total	37.9	21.0	21.6	30.7	17.0	52.9	18.1

Source: FEMA U. S. Fire Administration 2002
     Survey of the Needs of the US Fire Service

The above table breakdown is based on 7,159 departments reporting on Question 20. Departments were asked to circle all that apply, so departments could select multiple responses. Numbers may not add to totals due to rounding.

Q. 20: Which of the following programs or activities does your department conduct? Plans review; permit approval; routine testing of active systems (e.g., fire sprinkler, detection/alarm, smoke control); free distribution of home smoke alarms; juvenile firesetter program; school fire safety education program based on a national model curriculum; other prevention program.

54

Table 26
Who Conducts Fire-Code Inspections in the Community?
by Community Size
(Q. 21)

Population of Community	Full-Time Fire Department Inspectors	In-Service Firefighters	Building Department	Separate Inspection Department	Other	No One
1,000,000 or more	80.0%	30.0%	20.0%	10.0%	0.0%	0.0%
500,000 to 999,999	96.9	63.6	24.2	9.1	6.1	0.0
250,000 to 499,999	89.6	56.3	12.5	4.2	4.2	0.0
100,000 to 249,999	92.8	50.3	14.4	7.2	5.0	0.0
50,000 to 99,999	86.3	48.9	14.2	7.9	8.2	1.0
25,000 to 49,999	69.6	47.0	17.6	5.0	11.7	1.4
10,000 to 24,999	46.1	36.2	21.8	8.3	17.9	5.7
5,000 to 9,999	19.6	18.3	22.5	12.8	28.2	14.7
2,500 to 4,999	8.4	14.7	16.7	11.3	28.5	25.9
Under 2,500	3.3	10.9	12.1	11.8	22.1	39.4
Total	16.3	18.0	15.7	11.1	22.7	27.3

Source: FEMA U. S. Fire Administration 2002
Survey of the Needs of the US Fire Service

The above table breakdown is based on 8,218 departments reporting on Question 21. Departments were asked to circle all that apply, so departments could select multiple responses. Numbers may not add to totals due to rounding.

Q. 21: Who conducts fire code inspections in your community?

Table 27
Who Determines That a Fire Was Deliberately Set?
by Community Size
(Q. 22)

Population of Community	Fire Department Arson Investigator	Regional Arson Task Force Investigator	State Arson Investigator	Incident Commander	Police Department	Contract Investigator	Insurance Investigator	Other
1,000,000 or more	90.0%	0.0%	0.0%	10.0%	10.0%	0.0%	0.0%	10.0%
500,000 to 999,999	93.9	3.0	15.2	42.4	18.2	0.0	0.0	3.0
250,000 to 499,999	95.8	12.5	6.3	33.3	8.3	0.0	4.2	6.3
100,000 to 249,999	93.4	14.9	22.1	30.4	17.1	1.1	2.8	3.9
50,000 to 99,999	90.5	13.7	37.4	32.1	26.6	0.8	3.7	2.9
25,000 to 49,999	80.3	22.0	47.9	34.9	25.5	0.3	6.3	6.1
10,000 to 24,999	64.2	18.7	61.3	35.5	24.6	1.3	9.7	9.7
5,000 to 9,999	40.4	16.4	71.0	36.4	18.1	2.6	12.5	12.1
2,500 to 4,999	28.5	13.4	73.7	33.8	15.3	1.1	13.5	11.5
Under 2,500	14.5	10.8	73.3	29.3	14.1	1.3	15.1	9.7
Total	30.9	13.4	69.4	32.0	16.7	1.4	13.2	10.0

Source: FEMA U. S. Fire Administration 2002
Survey of the Needs of the US Fire Service

The above table breakdown is based on 8,376 departments reporting on Question 22. Departments were asked to circle all that apply, so departments could select multiple responses. Numbers may not add to totals due to rounding.

Q. 22: Who determines that a fire was deliberately set? "Incident commander" includes other first-in fire officer.

FACILITIES, APPARATUS AND EQUIPMENT

Fire Stations

Table 28 (p. 71) describes the average number of fire stations per department by size of community. Note that a community may have two or more fire stations, and each fire station may have two or more firefighting companies, each attached to a particular apparatus, such as an engine/pumper. Table 28 also describes the fraction of stations with characteristics that indicate potential needs, specifically age of station over 40 years, a lack of backup power, or a lack of exhaust emission control equipment. Table K converts these figures to total numbers of fire stations with those needs, by size of community and overall.

Table K. Number of Fire Stations With Characteristics Indicating Potential Need, by Size of Community Protected (Q. 23)

	Total Number of Fire Stations With Indicated Characteristics in Communities of This Population Size		
Population Protected	Over 40 Years Old	No Backup Power	Not Equipped for Exhaust Emission Control
1,000,000 or more	428	481	440
500,000 to 999,999	504	449	784
250,000 to 499,999	610	520	577
100,000 to 249,999	585	816	1,162
50,000 to 99,999	750	805	1,188
25,000 to 49,999	1,097	1,221	1,928
10,000 to 24,999	1,966	4,888	6,778
5,000 to 9,999	1,888	2,937	4,575
2,500 to 4,999	2,150	3,841	5,667
Under 2,500	5,516	11,564	14,789
Total	15,494	27,523	37,888
Percent of US total	32%	57%	78%

The above projections are based on 6,420 departments reporting on all four parts of Question 23. Numbers may not add to totals due to rounding. See Table 28.

Q. 23: Number of fire stations, number over 40 years old, number having backup power, number equipped for exhaust emission control (e.g., diesel exhaust extraction).

In addition to needs associated with the condition of fire stations, there are also questions about needs with respect to the number and coverage of fire stations. The number and coverage needed are those required to achieve response with sufficient fire suppression flow within a target period of time. The information contained in the Needs Assessment Survey is not sufficient to perform such a calculation, but a simplified version is possible.

The *Fire Suppression Rating Schedule* of the Insurance Services Office includes a number of guidelines and formulas to use in performing a complete assessment of the adequacy of fire department resources, but for this simplified calculation on adequacy of number of fire stations, Item 560 has a basis: "The built-upon area of the city should have a first-due engine company within 1- ½miles and a ladder -service company within 2- ½miles." [*] For this simplified calculation, we can use these two numbers as a range for the maximum distance from any point in the community to the nearest fire station.

NFPA 1710 states its requirements in terms of time, specifically, a requirement that 90% of responses by the initial arriving company shall be within 4 minutes. If the first-response area is considered as a circle with the fire station in the middle, and if emergency calls are evenly distributed throughout the response area, then 90% of responses will be within 95% of the distance from the fire station to the boundary of the response area.[**] If the average speed of fire apparatus is 21 mph, as it might be in the downtown area of a city, then the 4-minute requirement corresponds to a 1.5-mile requirement. If the average speed of fire apparatus is 36 mph, as it might be in a suburban or rural area, then the 4-minute requirement corresponds to a 2.5-mile requirement. In a very rural community, the average speed could be even higher, and the allowable distance would be even greater.

Note the limitations in this assumption: Item 560 implies that a larger maximum distance is acceptable for parts of the community that are not "built-upon"; this will be especially relevant for smaller communities. This larger maximum distance may or may not be on the order of the 2 ½miles cited for ladder -service companies responding in the built-upon area, so the use of 2 ½miles as an upper bound for calculation is done for convenience rather than through any compelling logic. Item 560 does not reflect variations in local travel speeds or the need for adequate fire flow by the responding apparatus; those issues are addressed elsewhere in the *Fire Suppression Rating Schedule*. This guideline is not a mandatory government requirement or a consensus voluntary standard.

To use this guideline with the data available from the Needs Assessment Survey, it is necessary to have a formula giving the maximum distance from fire station to any point in the community as a function of data collected in the survey. The Rand Institute developed such a formula for expected (i.e., average) distance as part of its extensive research on fire deployment issues in the 1960s and 1970s.[***]

---

[*] *Fire Suppression Rating Schedule*, New York: Insurance Services Office, Inc., August 1998, p. 28.

[**] If r is the distance from station to boundary, then the size of the response area is πr^2, and the radius of a circle with area equal to $0.9\pi r^2$ will be $r\sqrt{0.9}$ or approximately $0.95r$.

[***] Warren E. Walker, Jan M. Chaiken, and Edward J. Ignall, eds., *Fire Department Deployment Analysis*, Publications in Operations Research series of the Operations Research Society of America, New York: Elsevier North Holland, 1979, pp. 180-184.

The formula has been developed and tested against actual travel-distance data from selected fire departments for both straight-line travel and the more relevant right-angle travel that characterizes the grid layout of many communities. It has been developed assuming either a random distribution of fire stations throughout the community or an optimal placement of stations to minimize travel distances and times.

The formula is called the square root law: Expected distance = $k \sqrt{(A/n)}$
> where k is a proportionality constant
> > A is the community's area in square miles
> > n is the number of fire stations

Note the limitations of this approach, cited by the Rand authors: Most importantly, it ignores the effect of natural barriers, such as rivers or rail lines. It assumes an alarm is equally likely from any point in the community. It assumes a unit is always ready to respond from the nearest fire station.

If one further assumes that response areas can be approximated by circles with fire stations at the center, then expected distance equals one-half of maximum distance. If response areas are more irregularly shaped, expected distance will be a smaller fraction of maximum distance.

With these assumptions, the number of fire stations will be sufficient to provide acceptable coverage, defined as a maximum travel distance that is less than the ISO-based value, if the following is true:

$$A - \frac{1}{4}(n)(D_{max})^2/(k^2) < 0$$
where
> > A is the community's area in square miles
> > n is the number of fire stations
> > D_{max} is the maximum acceptable travel distance (1- ½ miles or 2 - ½ miles)
> > k is the Rand proportionality constant, which is assumed to be for right-
> > > angle travel and is 0.6267 for random station location and 0.4714 for optimal station location

Table L gives the estimates of need based on the four calculations (i.e., two possible maximums for travel distance times two possible location protocols for fire stations). It may be appropriate to use the shorter maximum distance for larger communities and the larger maximum distance for smaller communities. In fact, as noted, if the average speed achievable by fire apparatus is well above 36 mph, an even larger maximum distance is justified under NFPA 1710. Note also that NFPA 1720, the standard for volunteer fire departments, has no speed of response or distance requirement, reflecting the fact that very low population densities in the smallest communities mean the number of people exposed to long response times may be very small.

Also, while few if any communities will have optimal station locations, it is likely that most will have placements that are considerably better than random. If these two approaches are used, then Table L suggests that in every population interval, roughly two-thirds to three-fourths of fire departments have too few stations to provide the indicated coverage. (Specifically, if 1.5 miles is used for communities of 10,000 or more and 2.5 miles is used for smaller communities, with optimal location used for both, then Table L indicates that 65-76% of departments have too few stations, except for communities of 500,000 to 999,999 population, where the percentage is 82%.)

Table L. Estimated Percent of Fire Departments Lacking Sufficient Fire Stations to Achieve Specified Maximum Travel Distance by Size of Community Protected, Maximum Travel Distance Specified, and Assumption Regarding Optimality of Fire Station Placement (Q. 2, 23)

	Estimated Percent of Departments With Too Few Stations			
	Random station location		Optimal station location	
Population Protected	Maximum distance of 1.5 miles	Maximum distance of 2.5 miles	Maximum distance of 1.5 miles	Maximum distance of 2.5 miles
1,000,000 or more	80.0%	40.0%	70.0%	20.0%
500,000 to 999,999	87.9%	69.7%	81.8%	33.3%
250,000 to 499,999	83.0%	40.4%	63.8%	23.4%
100,000 to 249,999	92.1%	39.9%	66.9%	23.0%
50,000 to 99,999	88.0%	37.8%	64.4%	22.9%
25,000 to 49,999	88.7%	52.5%	70.4%	30.8%
10,000 to 24,999	91.6%	64.7%	75.7%	47.5%
5,000 to 9,999	93.3%	76.4%	82.2%	64.6%
2,500 to 4,999	93.5%	81.5%	85.8%	72.6%
Under 2,500	92.5%	80.8%	83.9%	72.6%

The above projections are based on 8,217 departments reporting on Questions 2 and 23.

Q. 2: Area (in square miles) your department has primary responsibility to protect (exclude mutual aid areas)
Q. 23: Number of fire stations

Remember the many limitations of this calculation procedure, however; a more complete calculation should be performed before drawing conclusions with regard to any particular community.

Apparatus

Table 29 (p. 72) characterizes the size of the engine/pumper fleet inventory, overall and by age of vehicle. Using the statistics from Table 2 on departments by population interval, one can identify the number of engines whose ages raise questions about the need for replacement. The breakdown by community size is shown in Figure 4 in terms of percent of apparatus and in Table M in terms of the number of apparatus.

Figure 4 indicates that in larger communities, those with at least 100,000 population, one-fifth to one-fourth (actually 20-24%) of engines are at least 15 years old. In smaller communities, those with less than 5,000 population, roughly one-half to two-thirds (actually 55-65%) of engines are at least 15 years old. Table M indicates there are more than 40,000 engines in use that are at least 15 years old, including more than 10,000 that are at least 30 years old. Most of these engines aged 15 years old or more are in use in smaller communities, with less than 5,000 population, but hundreds are in use in departments for every community size.

Vehicle age alone is not sufficient to confirm a need for replacement, but it is indicative of a potential need, which should be examined.

Table 29 also indicates the average number of ambulances or other patient transport vehicles per department, by community size. Communities of less than 10,000 population average less than one such vehicle per department; communities with 10,000 to 49,999 population average less than two; and even communities with less than 250,000 population average less than four. The averages are calculated over all departments, but larger shares of the small communities have fire departments that do not provide EMS, and this partially explains their lower numbers of ambulances per department.

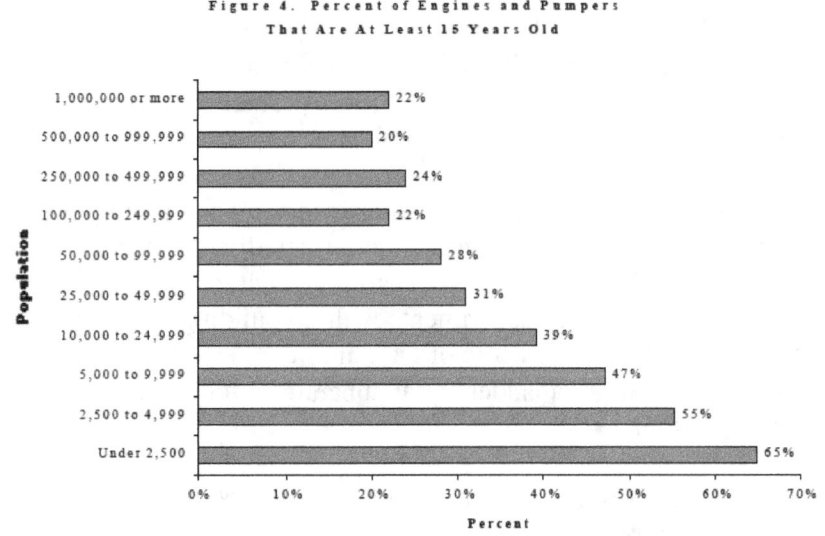

Figure 4. Percent of Engines and Pumpers
That Are At Least 15 Years Old

Table M. Number of Engines in Service, Limited to
Engines At Least 15 Years Old
by Age of Equipment and Size of Community Protected (Q. 24)

Population Protected	Total Number of Engines in Service of This Age in Fire Departments Protecting Communities of This Population Size		
	15 to 19 Years Old	20 to 29 Years Old	30+ Years Old
1,000,000 or more	159	101	0
500,000 to 999,999	268	91	7
250,000 to 499,999	282	79	47
100,000 to 249,999	441	181	43
50,000 to 99,999	560	282	102
25,000 to 49,999	863	569	190
10,000 to 24,999	1,791	1,820	426
5,000 to 9,999	2,032	2,214	944
2,500 to 4,999	2,012	3,063	1,646
Under 2,500	4,838	8,602	6,854
Total	13,247	17,000	10,259
Percent of US total	16%	21%	13%

The above projections are based on 4,769 departments reporting on all parts of Question 24. Numbers may not add to totals due to rounding. See Table 29.

Q. 24: Number of engines/pumpers in service. Total, 0-14 years old, 15-19 years old, 20-29 years old, 30 or more years old, unknown age

Table 30 (p. 73) provides information on the percentage of departments with ladder/aerial apparatus. This type of apparatus is of use for buildings at least four stories in height, although it can also be used for shorter buildings with access problems for ground ladders.

Therefore, it is useful to compare the percentage of departments, by community size, having no ladder/aerial apparatus with the percentage having buildings 4 stories high or higher. (See Table 31, p. 74.) If the percentage of departments without ladder/aerial apparatus is greater than the percentage of departments with no buildings of at least 4 stories in height, then the difference is a measure of the minimum percentage of departments that could justify acquiring a ladder/aerial apparatus but do not have one. Table N provides that comparison.

Table N indicates that at least 2.1% of departments (2.1% minus 0.0%) protecting communities of 250,000 to 499,999 population have no ladder/aerial apparatus but have at least one building tall enough to justify such apparatus. This is also true for 1.1% of departments (3.4% minus 2.3%) protecting communities of 100,000 to 249,999 population; 0.5% of departments (9.3% minus 8.8%) protecting communities of 50,000 to 99,999 population; 2.2% of departments (32.2% minus 29.8%) protecting communities of 10,000 to 24,999 population; 10.3% of departments (61.9% minus 51.6%) protecting communities of 5,000 to 9,999 population; 16.6% of departments (86.0% minus 69.4%) protecting communities of 2,500 to 4,999 population; and 15.4% of departments (95.7% minus 80.3%) protecting communities of less than 2,500 population.

Table N. Departments With No Ladder/Aerial Apparatus vs. Departments With No Buildings of At Least 4 Stories in Height Percent of Departments, by Size of Community Protected (Q. 25)

Population Protected	No Ladder/Aerial Apparatus	No Buildings At Least 4 Stories in Height
1,000,000 or more	0.0%	0.0%
500,000 to 999,999	0.0%	0.0%
250,000 to 499,999	2.1%	0.0%
100,000 to 249,999	3.4%	2.3%
50,000 to 99,999	9.3%	8.8%
25,000 to 49,999	13.7%	15.1%
10,000 to 24,999	32.2%	29.8%
5,000 to 9,999	61.9%	51.6%
2,500 to 4,999	86.0%	69.4%
Under 2,500	95.7%	80.3%
Total	76.5%	64.1%

The above projections are based on 7,216 departments reporting on the first part of Question 25 and 7,082 reporting on the second part. See Tables 30-31.

Q. 25: Number of ladders/aerials in service. Number of buildings in community that are 4 or more stories in height. None, 1-5, 6-10, 11 or more

Personal Protective Equipment and Clothing

Table 32 (p. 75) indicates what percentage of emergency responders on a single shift are equipped with portable radios. Tables 33 and 34 (pp. 76-77) indicate what fractions of those radios are water-resistant and intrinsically safe in an explosive atmosphere, respectively. Finally, Table 35 (p. 78) indicates whether departments have reserve radios at least equal to 10% of the in-service radios.

Figure 5 and Table O translates the results of Tables 32-34 into estimated percentages of emergency responders on a shift who lack radios and estimated percentages of radios that lack water-resistance or intrinsic safety in an explosive atmosphere.

For communities of 1 million population, an estimated 18% of emergency responders on a shift lack radios. For communities of 10,000 to under 1 million population, roughly one-fourth (24-29%) of emergency responders on a shift are estimated to lack radios.

The percentage without radios increases as community size decreases, reaching a majority of emergency responders on a shift (51%) lacking radios for communities with less than 2,500 population.

Table 35 is considered to speak for itself, without conversion.

In the last two columns of Table O, the range indicates two approaches to the "Don't Know" responses. The higher numbers assume that the "Don't Know" responses are all cases of unrecognized need, because one would expect fire department managements to be aware if their radios have these sophisticated features. The lower numbers use the more conventional assumption that "Don't Know" respondents look like the other respondents, so the former are statistically allocated over the latter.

Overall, roughly three-fifths of radios are not water-resistant and two-thirds are not intrinsically safe in an explosive atmosphere.

From Table 35, two-fifths to half of departments in communities with at least 50,000 population have sufficient reserve radios to replace at least 10% of in-service radios. This fraction falls with community size, reaching one-ninth for communities with less than 2,500 population.

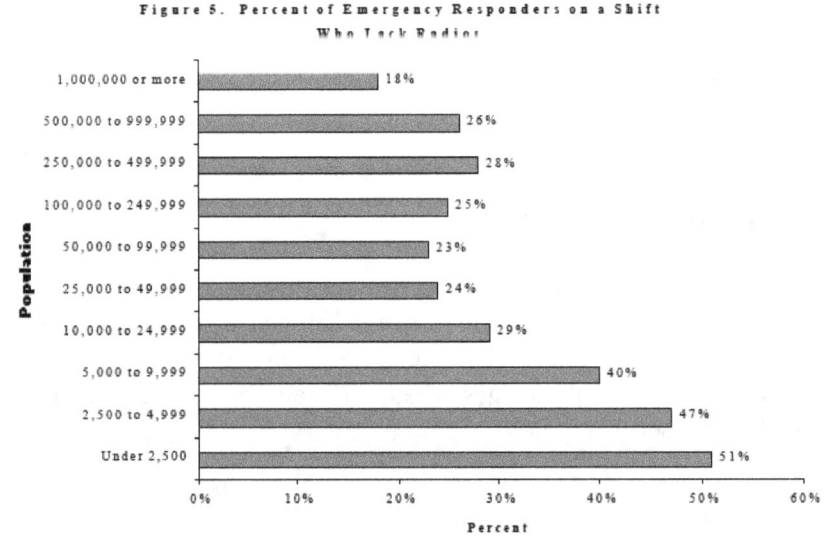

Figure 5. Percent of Emergency Responders on a Shift Who Lack Radios

64

**Table O. Emergency Responders on a Shift Who Lack Radios
and Radios Lacking Water-Resistance or Intrinsic Safety
in an Explosive Atmosphere
by Size of Community Protected (Q. 27a, 27b, 27c)**

Population Protected	Percent of Emergency Responders on Shift Lacking Radios	Percent of Radios Lacking	
		Water Resistance	Intrinsic Safety in Explosive Atmosphere
1,000,000 or more	18%	14-21%	36-46%
500,000 to 999,999	26%	38-40%	31-39%
250,000 to 499,999	28%	37-38%	48-53%
100,000 to 249,999	25%	45-48%	45-53%
50,000 to 99,999	23%	32-40%	38-48%
25,000 to 49,999	24%	38-45%	44-53%
10,000 to 24,999	29%	41-48%	51-61%
5,000 to 9,999	40%	50-57%	58-70%
2,500 to 4,999	47%	57-64%	64-77%
Under 2,500	51%	64-72%	69-84%
Total	45%	56-64%	61-76%

The above projections are based on 8,343 departments reporting on Question 27a, 8,307 reporting on Question 27b, and 8,269 reporting on Question 27c. "Most" and "Some" are converted to 2/3 and 1/3. See Tables 32-34.

Q. 27a: How many of your emergency responders on-duty on a single shift can be equipped with portable radios? All, Most, Some, None
Q. 27b: How many of your portable radios are water-resistant? All, Most, Some, None
Q. 27c: How many of your portable radios are intrinsically safe in an explosive atmosphere? All, Most, Some, None

Table 36 (p. 79) estimates how many emergency responders on a shift or otherwise on-duty are equipped with self-contained breathing apparatus (SCBA).

Table 37 (p. 80) estimates what fraction of the SCBA units are at least 10 years old.

The breakdown of need by community size is given in Figure 6 and Table P, in terms of percent of personnel on a shift who lack SCBA and percent of SCBA units that are at least 10 years old, both by size of community protected.

The second-column percentages in Table P cannot be safely converted to number of needed SCBA units using Table A, because the need is for sufficient units to equip all personnel on a shift, and so the base required to apply the percentages is the number of personnel per shift, not the number per department, as given in Table A. This base is not known.

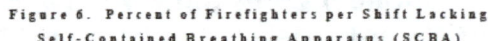

Figure 6. Percent of Firefighters per Shift Lacking
Self-Contained Breathing Apparatus (SCBA)

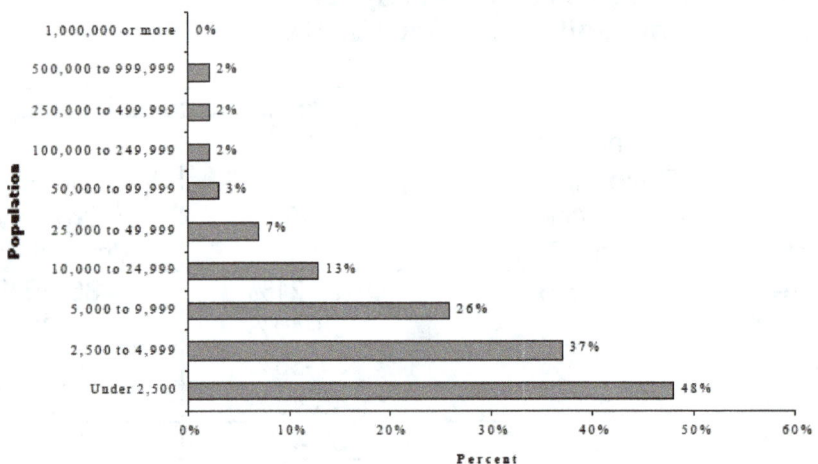

Table P. Firefighters per Shift Lacking SCBA and
SCBA Units At Least 10 Years Old, by Size of Community (Q. 28a, 28b)

Population Protected	Estimated Percent of Firefighters per Shift Not Equipped With SCBA	Estimated Percent of SCBA Units That Are At Least 10 Years Old
1,000,000 or more	0%	26%
500,000 to 999,999	2%	13%
250,000 to 499,999	2%	21%
100,000 to 249,999	2%	24%
50,000 to 99,999	3%	25%
25,000 to 49,999	7%	27%
10,000 to 24,000	13%	33%
5,000 to 9,999	26%	38%
2,500 to 4,999	37%	41%
Under 2,500	48%	53%
Total	36%	45%

The above projections are based on 8,346 departments reporting on Question 28a and 8,303 reporting on Question 28b. "Most" and "Some" are converted to 2/3 and 1/3. "Don't Know" responses to Question 28b are proportionally allocated. See Tables 36-37.

Q. 28a: How many emergency responders on-duty on a single shift can be equipped with self-contained breathing apparatus (SCBA)? All, Most, Some, None
Q. 28b: How many of your SCBA are 10 years old or older? All, Most, Some, None

For communities with at least 50,000 population, at most 3% of emergency responders on a shift in the average department need SCBA units. This rises to roughly half needing SCBA units in the average department protecting communities with less than 2,500 population.

For larger communities, roughly one-fourth of SCBA units are at least 10 years old, while for smaller communities, the fraction rises to one-half.

Table 38 (p. 81) indicates what fraction of emergency responders on a single shift are equipped with Personal Alert Safety System (PASS) devices.

The breakdown of need is given in Figure 7 and Table Q, in terms of percent of personnel on a shift who lack PASS devices, by size of community protected.

These percentages cannot be safely converted to number of needed PASS devices, using Table A, because the need is for sufficient units to equip all personnel on a shift, and so the base required to apply the percentages is the number of personnel per shift, not the number per department. This base is not known.

For communities with at least 50,000 population, at most 3% of emergency responders on a shift in the average department need PASS devices.

This rises to one-eighth for communities with 10,000 to 24,999 population, one-fourth for communities with 5,000 to 9,999 population, two-fifths for communities with 2,500 to 4,999 population, and roughly three-fifths needing PASS units in the average department protecting communities with less than 2,500 population.

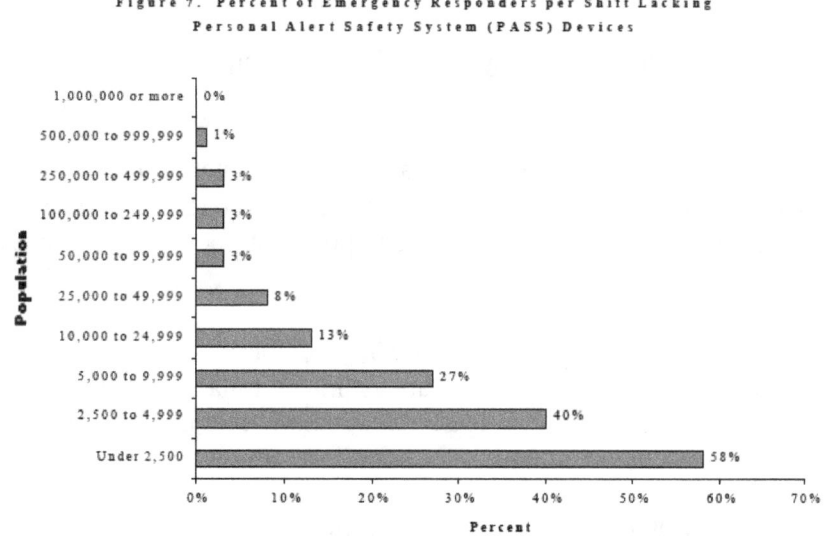

Figure 7. Percent of Emergency Responders per Shift Lacking Personal Alert Safety System (PASS) Devices

Table Q. Estimated Average Percent of Emergency Responders per Shift Not Provided With PASS Devices, by Size of Community (Q. 29)

Population Protected	Emergency Responders per Shift Not Provided with PASS Devices
1,000,000 or more	0%
500,000 to 999,999	1%
250,000 to 499,999	3%
100,000 to 249,999	3%
50,000 to 99,999	3%
25,000 to 49,999	8%
10,000 to 24,999	13%
5,000 to 9,999	27%
2,500 to 4,999	40%
Under 2,500	58%
Total	42%

The above projections are based on 8,326 departments reporting on Question 29. "Most" and "Some" are converted to 2/3 and 1/3. See Table 38.

Q. 29: How many of your emergency responders on-duty on a single shift are equipped with Personal Alert Safety System (PASS) devices? All, Most, Some, None

Table 39 (p. 82) indicate how many emergency responders are equipped with their own personal protective clothing.

Here, the percentages can be applied to Table A, because all firefighters should have their own personal protective clothing. Using the simple estimation approach established earlier, it is estimated that 57,000 firefighters are not equipped with personal protective clothing.

The breakdown by community size is shown in Figure 8 and Table R. For communities with at least 5,000 population, 0-3% of firefighters are estimated to lack personal protective clothing. For communities, with at least 500,000 population, none are estimated to lack personal protective clothing.

For communities of 2,500 to 4,999 population, 5% of firefighters are estimated to lack personal protective clothing. For communities of less than 2,500 population, the percentage is 10%.

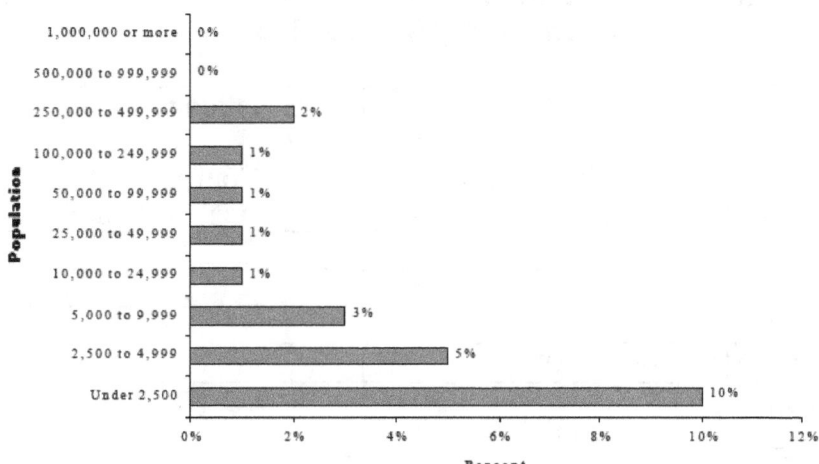

Nearly all of the firefighters estimated to lack personal protective clothing are in fire departments that protect communities with less than 10,000 population.

More specifically, communities with less than 2,500 population are estimated to account for 42,000 of the 57,000 unprotected firefighters (74% of the total). Communities with 2,500 to 4,999 population are estimated to account for 8,000 of the 57,000 unprotected firefighters (13% of the total). And communities with 5,000 to 9,999 population are estimated to account for 4,000 of the 57,000 unprotected firefighters (7% of the total).

The total for communities of less than 10,000 population is 95% of the total number of unprotected firefighters.

Except for odd results for communities of 250,000 to 499,999, in the average department protecting a community with at least 25,000 population, roughly one-sixth of the personal protective clothing is at least 10 years old. (See Table 40, p. 83.) This rises to nearly half for communities with less than 2,500 population.

Table 41 (p. 84) describes the extent to which departments have reserve protective clothing sufficient to equip 10% of responders.

Apart from some volatile results for communities of 500,000 or more population, two-thirds of the departments protecting communities with at least 100,000 population have reserve clothing sufficient to equip 10% of responders.

This falls to three-fifths for communities of 25,000 to 99,999 population and falls further as community size shrinks, reaching one-third for communities with less than 2,500 population.

**Table R. Firefighters in Departments Where Not All Firefighters
Are Equipped With Personal Protective Clothing and
Percent of Personal Protective Clothing That Is At Least 10 Years Old
by Size of Community (Q. 30a, 30b)**

Population Protected	Estimated Firefighters Lacking Personal Protective Clothing	Estimated Percent of Personal Protective Clothing That Is At Least 10 Years Old
1,000,000 or more	0	7%
500,000 to 999,999	0	16%
250,000 to 499,999	0*	48%
100,000 to 249,999	0*	17-18%
50,000 to 99,999	0*	15-16%
25,000 to 49,999	1,000	19%
10,000 to 24,999	1,000	24%
5,000 to 9,999	4,000	30-31%
2,500 to 4,999	8,000	36%
Under 2,500	42,000	45%
Total	57,000	37-38%

* Rounds to zero but is not zero

The above projections are based on 8,394 departments reporting on Question 30a and 8,364 reporting on Question 30b. "Most" and "Some" are converted to 2/3 and 1/3. Ranges in the last column reflect the difference between allocating "Don't Know" responses proportionally and treating them all as unmet need. Numbers are shown to nearest 1,000 and may not sum to totals because of rounding. See Tables 39-40.

Q. 30a: How many of your emergency responders are equipped with personal protective clothing? All, Most, Some, None
Q. 30b: How much of your personal protective clothing is at least 10 years old? All, Most, Some, None

Table 28
Number of Fire Stations and Selected Characteristics
by Community Size
(Q. 23)

Population of Community	Average Number of Stations	Percent Stations Over 40 Years Old	Percent Stations Having Backup Power	Percent Stations Equipped for Exhaust Control
1,000,000 or more	75.80	43.4%	51.2%	55.3%
500,000 to 999,999	38.91	34.1	69.6	47.0
250,000 to 499,999	20.53	46.4	60.4	56.1
100,000 to 249,999	10.72	25.4	64.6	49.6
50,000 to 99,999	5.35	28.8	69.1	54.4
25,000 to 49,999	3.36	31.0	65.5	45.5
10,000 to 24,999	2.98	23.2	42.3	20.0
5,000 to 9,999	1.53	34.0	47.1	17.6
2,500 to 4,999	1.34	35.1	37.3	7.5
Under 2,500	1.20	34.2	28.3	8.3
Total	1.84	31.9	43.3	21.9

Source: FEMA US Fire Administration 2002
Survey of the Needs of the US Fire Service

The above projections are based on 6,420 departments answering all four parts of Question 23. Numbers may not add to totals due to rounding.

Q. 23: Number of fire stations, number over 40 years old, number having backup power, number equipped for exhaust emission control (e.g., diesel exhaust extraction).

Table 29
Average Number of Engines/Pumpers and Ambulances* in Service
and Age of Engine/Pumper Apparatus
by Community Size
(Q. 24, 26)

Population of Community	Average Number of Engines	Engines 0-14 Years Old	Engines 15-19 Years Old	Engines 20-29 Years Old	Engines 30 or More Years Old	Average Number of Ambulances*
1,000,000 or more	91.11	71.11	12.22	7.78	0.00	23.86
500,000 to 999,999	47.10	38.03	7.06	2.39	0.19	15.58
250,000 to 499,999	26.96	20.59	4.41	1.23	0.73	6.69
100,000 to 249,999	13.75	10.63	2.05	0.84	0.20	3.75
50,000 to 99,999	6.94	4.97	1.15	0.58	0.21	2.47
25,000 to 49,999	4.99	3.38	0.82	0.54	0.18	1.71
10,000 to 24,999	3.67	2.23	0.63	0.64	0.15	1.05
5,000 to 9,999	3.06	1.63	0.56	0.61	0.26	0.65
2,500 to 4,999	2.69	1.21	0.44	0.67	0.36	0.44
Under 2,500	2.32	0.82	0.36	0.64	0.51	0.32
Total	3.09	1.55	0.50	0.65	0.39	0.64

* "Ambulances" include other patient transport vehicles.

Source: FEMA US Fire Administration 2002
　　　　Survey of the Needs of the US Fire Service

The above table breakdown, except for the last column, is based on 4,769 departments answering all parts of Question 24. The last column is based on 7,968 departments answering Question 26.

Numbers may not add to totals due to rounding.

Q. 24: Number of engines/pumpers in service, number 0-14 years old, number 15-19 years old, number 20-29 years old, number 30 or more years old, number unknown age.
Q. 26: Number of ambulances or other patient transport vehicles.

Table 30
Number of Ladders/Aerials In-Service, by Community Size
(Q. 25)

For Departments Protecting Populations of 250,000 or More, Percent of Departments With

Population of Community	No Ladders/ Aerials	1-5 Ladders/ Aerials	6-9 Ladders/ Aerials	10-19 Ladders/ Aerials	20 or More Aerials/ Ladders	Total
1,000,000 or more	0.0%	0.0%	0.0%	30.0%	70.0%	100.0%
500,000 to 999,999	0.0	15.7	28.1	46.9	9.4	100.0%
250,000 to 499,999	2.1	59.6	19.1	14.9	4.3	100.0%

For Departments Protecting Populations of Less Than 250,000, Percent of Departments With

Population of Community	No Ladders/ Aerials	1 Ladder/ Aerial	2 Ladders/ Aerials	3-4 Ladders/ Aerials	5 or More Ladders/ Aerials	Total
100,000 to 249,999	3.4%	21.6%	25.0%	31.3%	18.8%	100.0%
50,000 to 99,999	9.3	39.6	33.9	15.8	1.4	100.0%
25,000 to 49,999	13.7	59.4	23.3	3.6	0.0	100.0%
10,000 to 24,999	32.2	60.1	7.2	0.5	0.0	100.0%
5,000 to 9,999	61.9	36.2	1.7	0.0	0.0	100.0%
2,500 to 4,999	86.0	13.5	0.5	0.0	0.0	100.0%
Under 2,500	95.7	4.0	0.3	0.0	0.0	100.0%

Source: FEMA US Fire Administration 2002
  Survey of the Needs of the US Fire Service

The above table breakdown is based on 7,216 departments reporting on Question 25. Numbers may not add to totals due to rounding.

Table 31
Number of Buildings in Community That Are 4 or More Stories in Height by Community Size
(Q. 25)

Number of Buildings

Population of Community	None		1 to 5		6 to 9		10 or more		Total	
	Number Depts	Percent	Number Depts	Percent	Number Depts	Percent	Number Depts	Percent	Number Depts	Percent
1,000,000 or more	0	0.0%	0	0.0%	0	0.0%	13	100.0%	13	100.0%
500,000 to 999,999	0	0.0	0	0.0	1	2.6	37	97.4	38	100.0
250,000 to 499,999	0	0.0	0	0.0	2	3.1	62	96.9	64	100.0
100,000 to 249,999	5	2.3	27	12.6	20	9.3	163	75.8	215	100.0
50,000 to 99,999	43	8.8	100	20.5	71	14.6	273	56.1	487	100.0
25,000 to 49,999	159	15.1	316	30.0	202	19.2	377	35.8	1,053	100.0
10,000 to 24,999	847	29.8	1,004	35.3	562	19.8	430	15.2	2,843	100.0
5,000 to 9,999	1,871	51.6	1,146	31.6	355	9.8	258	7.1	3,629	100.0
2,500 to 4,999	3,173	69.4	1,083	23.7	201	4.4	115	2.5	4,572	100.0
Under 2,500	10,794	80.3	2,337	17.4	200	1.5	109	0.8	13,440	100.0
Total	16,893	64.1	6,013	22.8	1,611	6.1	1,836	7.0	26,354	100.0

Source: FEMA US Fire Administration 2002
Survey of the Needs of the US Fire Service

The above projections are based on 7,082 departments reporting on Question 25. Numbers may not add to totals due to rounding.

74

Table 32
How Many of Department's Emergency Responders
on a Single Shift Are Equipped With Portable Radios?
by Community Size
(Q. 27a)

Population of Community	All		Most		Some		None		Total	
	Number Depts	Percent	Number Depts	Percent	Number Depts	Percent	Number Depts	Percent	Number Depts	Percent
1,000,000 or more	9	69.2%	1	7.7%	3	23.1%	0	0.0%	13	100.0%
500,000 to 999,999	16	42.1	14	36.8	8	21.1	0	0.0	38	100.0
250,000 to 499,999	27	42.2	20	31.3	17	26.6	0	0.0	64	100.0
100,000 to 249,999	99	46.0	68	31.6	48	22.3	0	0.0	215	100.0
50,000 to 99,999	233	47.8	171	35.1	83	17.0	0	0.0	487	100.0
25,000 to 49,999	508	48.2	329	31.3	211	20.0	5	0.5	1,053	100.0
10,000 to 24,999	1,217	42.8	754	26.5	858	30.2	14	0.5	2,843	100.0
5,000 to 9,999	973	26.8	950	26.2	1,688	46.5	18	0.5	3,629	100.0
2,500 to 4,999	846	18.5	1,074	23.5	2,589	56.6	64	1.4	4,572	100.0
Under 2,500	2,099	15.6	3,010	22.4	7,456	55.5	875	6.5	13,440	100.0
Total	6,026	22.9	6,392	24.3	12,959	49.2	976	3.6	26,354	100.0

Source: FEMA US Fire Administration 2002
Survey of the Needs of the US Fire Service

The above projections are based on 8,343 departments reporting on Question 27a. Numbers may not add to totals due to rounding.

Q. 27a: How many of your emergency responders on-duty on a single shift can be equipped with portable radios?

75

Table 33
What Fraction of Department's Portable Radios Are Water-Resistant?
by Community Size
(Q. 27b)

Population of Community	All		Most		Some		None		Don't Know		Total	
	Number Depts	Percent	Number Depts	Percent	Number Depts	Percent	Number Depts	Percent	Number Depts	Percent	Number Depts	Percent
1,000,000 or more	10	77.0%	0	0.0%	1	7.7%	1	7.7%	1	7.7%	13	100.0%
500,000 to 999,999	15	39.5	8	21.1	8	21.1	6	15.8	1	2.6	38	100.0
250,000 to 499,999	31	48.4	8	12.5	11	17.2	13	20.3	1	1.6	64	100.0
100,000 to 249,999	85	39.5	27	12.6	23	10.7	67	31.1	13	6.0	215	100.0
50,000 to 99,999	232	47.6	68	14.0	52	10.7	83	17.0	52	10.7	487	100.0
25,000 to 49,999	441	42.1	149	14.2	116	11.1	227	21.7	115	10.9	1,048	100.0
10,000 to 24,999	994	35.1	530	18.7	349	12.3	614	21.7	343	12.1	2,829	100.0
5,000 to 9,999	968	26.8	574	15.9	597	16.5	973	27.0	498	13.8	3,611	100.0
2,500 to 4,999	888	19.7	715	15.9	755	16.8	1,401	31.1	750	16.6	4,508	100.0
Under 2,500	2,145	17.1	1,352	10.8	1,962	15.6	4,173	33.2	2,928	23.3	12,565	100.0
Total	5,808	22.9	3,431	13.5	3,875	14.3	7,560	29.8	4,705	18.5	25,378	100.0

Source: FEMA US Fire Administration 2002
Survey of the Needs of the US Fire Service

The above projections are based on 8,307 departments reporting on Question 27b. Numbers may not add to totals due to rounding.

Q. 27b: How many of your portable radios are water-resistant?

76

Table 34
What Fraction of Department's Portable Radios
Are Intrinsically Safe in an Explosive Atmosphere?
by Community Size
(Q. 27c)

Population of Community	All		Most		Some		None		Don't Know		Total	
	Number Depts	Percent	Number Depts	Percent	Number Depts	Percent	Number Depts	Percent	Number Depts	Percent	Number Depts	Percent
1,000,000 or more	4	30.8%	3	23.1%	3	23.1%	1	7.7%	2	15.4%	13	100.0%
500,000 to 999,999	18	47.4	5	13.2	6	15.8	5	13.2	4	10.5	38	100.0
250,000 to 499,999	18	28.1	11	17.2	15	23.4	14	21.8	6	9.4	64	100.0
100,000 to 249,999	77	35.8	22	10.2	33	15.3	53	24.7	30	14.4	215	100.0
50,000 to 99,999	199	40.9	48	9.9	61	12.5	100	20.5	79	16.2	487	100.0
25,000 to 49,999	360	34.4	134	12.8	141	13.4	240	22.9	173	16.5	1,048	100.0
10,000 to 24,999	737	26.0	337	11.9	404	14.3	737	26.0	615	21.7	2,829	100.0
5,000 to 9,999	691	19.1	367	10.1	489	13.6	1,049	29.1	1,015	28.1	3,611	100.0
2,500 to 4,999	621	13.8	366	8.1	566	12.6	1,361	30.2	1,594	35.4	4,508	100.0
Under 2,500	1,210	9.6	745	5.9	1,174	9.3	3,243	25.8	6,194	49.3	12,565	100.0
Total	3,935	15.5	2,035	8.0	2,891	11.4	6,805	26.8	9,712	38.3	25,378	100.0

Source: FEMA US Fire Administration 2002
Survey of the Needs of the US Fire Service

The above projections are based on 8,269 departments reporting on Question 27c. Numbers may not add to totals due to rounding.

Q. 27c: How many of your portable radios are intrinsically safe in an explosive atmosphere?

Table 35
Does Department Have Reserve Portable Radios
Equal to or Greater Than 10% of In-Service Radios?
by Community Size
(Q. 27d)

	Yes		No		Don't Know		Total	
Population of Community	Number Depts	Percent	Number Depts	Percent	Number Depts	Percent	Number Depts	Percent
1,000,000 or more	6	46.2%	6	46.2%	1	7.6%	13	100.0%
500,000 to 999,999	17	44.7	17	44.7	4	10.5	38	100.0
250,000 to 499,999	26	40.6	34	53.1	4	6.3	64	100.0
100,000 to 249,999	95	44.2	116	54.0	4	1.8	215	100.0
50,000 to 99,999	207	42.5	272	55.9	8	1.6	487	100.0
25,000 to 49,999	390	37.3	646	61.6	12	1.1	1,048	100.0
10,000 to 24,999	863	30.5	1,915	67.7	52	1.8	2,829	100.0
5,000 to 9,999	723	20.0	2,829	78.4	58	1.6	3,611	100.0
2,500 to 4,999	727	16.1	3,660	81.2	122	2.7	4,508	100.0
Under 2,500	1,486	11.8	10,544	83.9	540	4.3	12,565	100.0
Total	4,538	17.9	20,035	78.9	804	3.2	25,378	100.0

Source: FEMA US Fire Administration 2002
        Survey of the Needs of the US Fire Service

The above projections are based on 8,288 departments reporting on Question 27d. Numbers may not add to totals due to rounding.

Q. 27d: Do you have reserve portable radios equal to or greater than 10% of your in-service radios?

Table 36
How Many Emergency Responders on a Single Shift Are Equipped With Self-Contained Breathing Apparatus (SCBA)?
by Community Size
(Q. 28a)

Population of Community	All		Most		Some		None		Total	
	Number Depts	Percent	Number Depts	Percent	Number Depts	Percent	Number Depts	Percent	Number Depts	Percent
1,000,000 or more	13	100.0%	0	0.0%	0	0.0%	0	0.0%	13	100.0%
500,000 to 999,999	36	94.7	2	5.3	0	0.0	0	0.0	38	100.0
250,000 to 499,999	61	95.3	3	4.7	0	0.0	0	0.0	64	100.0
100,000 to 249,999	204	94.8	7	3.4	4	1.7	0	0.0	215	100.0
50,000 to 99,999	448	92.0	31	6.3	8	1.7	0	0.0	487	100.0
25,000 to 49,999	877	83.3	145	13.8	31	2.9	0	0.0	1,053	100.0
10,000 to 24,999	1,925	67.7	729	25.6	186	6.6	2	0.1	2,843	100.0
5,000 to 9,999	1,502	41.4	1,473	40.6	651	17.9	3	0.1	3,629	100.0
2,500 to 4,999	1,105	24.2	1,938	42.4	1,503	32.9	26	0.6	4,572	100.0
Under 2,500	1,750	13.0	4,451	33.1	6,945	51.7	293	2.2	13,440	100.0
Total	7,920	30.0	8,783	33.3	9,327	35.3	324	1.2	26,354	100.0

Source: FEMA US Fire Administration 2002
Survey of the Needs of the US Fire Service

The above projections are based on 8,346 departments reporting on Question 28a. Numbers may not add to totals due to rounding.

Q. 28a: How many emergency responders on-duty on a single shift can be equipped with self-contained breathing apparatus (SCBA)?

Table 37
How Much of Department's
SCBA Equipment Is At Least 10 Years Old?
by Community Size
(Q. 28b)

Population of Community	All Number Depts	All Percent	Most Number Depts	Most Percent	Some Number Depts	Some Percent	None Number Depts	None Percent	Don't Know Number Depts	Don't Know Percent	Total Number Depts	Total Percent
1,000,000 or more	0	0.0%	1	10.0%	3	20.0%	8	60.0%	1	10.0%	13	100.0%
500,000 to 999,999	0	0.0	3	7.9	6	15.8	28	73.7	1	2.6	38	100.0
250,000 to 499,999	2	3.1	8	12.5	11	17.2	41	64.0	2	3.1	64	100.0
100,000 to 249,999	13	6.0	22	10.2	72	33.4	108	50.2	1	0.6	215	100.0
50,000 to 99,999	32	6.6	50	10.2	145	29.8	252	51.7	8	1.6	487	100.0
25,000 to 49,999	82	7.8	109	10.4	361	34.3	492	46.8	9	0.8	1,053	100.0
10,000 to 24,999	253	8.9	498	17.5	967	34.0	1,109	39.0	16	0.6	2,843	100.0
5,000 to 9,999	422	11.6	689	19.0	1,359	37.4	1,115	30.7	44	1.2	3,629	100.0
2,500 to 4,999	603	13.2	958	20.9	1,803	39.4	1,177	25.7	32	0.7	4,572	100.0
Under 2,500	3,272	24.3	2,940	21.9	4,507	33.5	2,460	18.3	260	1.9	13,440	100.0
Total	4,682	17.8	5,280	20.0	9,237	35.0	6,782	25.7	373	1.4	26,354	100.0

Source: FEMA US Fire Administration 2002
Survey of the Needs of the US Fire Service

The above projections are based on 8,303 departments reporting on Question 28b. Numbers may not add to totals due to rounding.

Q. 28b: How many of your self-contained breathing apparatus (SCBA) are 10 years old or older?

Table 38
What Fraction of Emergency Responders on a Single Shift Are Equipped With Personal Alert Safety System (PASS) Devices?
by Community Size
(Q. 29)

Population of Community	All		Most		Some		None		Total	
	Number of Depts	Percent	Number of Depts	Percent	Number of Depts	Percent	Number of Depts	Percent	Number of Depts	Percent
1,000,000 or more	13	100.0%	0	0.0%	0	0.0%	0	0.0%	13	100.0%
500,000 to 999,999	37	97.4	1	2.6	0	0.0	0	0.0	38	100.0
250,000 to 499,999	61	95.3	2	3.2	0	0.0	1	1.6	64	100.0
100,000 to 249,999	206	96.0	2	1.1	4	1.7	3	1.2	215	100.0
50,000 to 99,999	450	92.3	27	5.5	9	1.9	1	0.3	487	100.0
25,000 to 49,999	902	85.6	90	8.5	36	3.4	25	2.4	1,053	100.0
10,000 to 24,999	2,112	74.3	417	14.7	233	8.2	81	2.9	2,843	100.0
5,000 to 9,999	1,836	50.6	898	24.7	629	17.3	266	7.3	3,629	100.0
2,500 to 4,999	1,578	34.5	1,120	24.5	1,244	27.2	631	13.8	4,572	100.0
Under 2,500	2,827	21.0	2,352	17.5	3,916	29.1	4,345	32.3	13,440	100.0
Total	10,018	38.0	4,909	18.6	6,073	23.0	5,353	20.3	26,354	100.0

Source: FEMA US Fire Administration 2002
Survey of the Needs of the US Fire Service

The above projections are based on 8,326 departments reporting on Question 29. Numbers may not add to totals due to rounding.

Q. 29: How many of your emergency responders on-duty on a single shift are equipped with Personal Alert Safety System (PASS) devices?

81

Table 39
What Fraction of Emergency Responders
Are Equipped With Personal Protective Clothing?
by Community Size
(Q. 30a)

Population of Community	All		Most		Some		None		Total	
	Number Depts	Percent	Number Depts	Percent	Number Depts	Percent	Number Depts	Percent	Number Depts	Percent
1,000,000 or more	13	100.0%	0	0.0%	0	0.0%	0	0.0%	13	100.0%
500,000 to 999,999	38	100.0	0	0.0	0	0.0	0	0.0	38	100.0
250,000 to 499,999	61	95.3	3	4.7	0	0.0	0	0.0	64	100.0
100,000 to 249,999	211	98.2	4	1.8	0	0.0	0	0.0	215	100.0
50,000 to 99,999	480	98.6	6	1.2	0	0.0	1	0.3	487	100.0
25,000 to 49,999	1,036	98.4	14	1.3	2	0.2	2	0.2	1,053	100.0
10,000 to 24,999	2,754	96.9	81	2.8	6	0.2	2	0.1	2,843	100.0
5,000 to 9,999	3,302	91.0	294	8.1	26	0.7	8	0.2	3,629	100.0
2,500 to 4,999	4,050	88.6	413	9.0	97	2.1	11	0.3	4,572	100.0
Under 2,500	10,372	77.2	2,174	16.2	803	6.0	90	0.7	13,440	100.0
Total	22,319	84.7	2,986	11.3	934	3.5	115	0.4	26,354	100.0

Source: FEMA US Fire Administration 2002
Survey of the Needs of the US Fire Service

The above projections are based on 8,394 departments reporting on Question 30a. Numbers may not add to totals due to rounding.

Q. 30a: How many of your emergency responders are equipped with personal protective clothing?

82

Table 40
How Much of Department's Personal
Protective Clothing Is At Least 10 Years Old?
by Community Size
(Q. 30b)

Population of Community	All		Most		Some		None		Don't Know		Total	
	Number Depts	Percent	Number Depts	Percent	Number Depts	Percent	Number Depts	Percent	Number Depts	Percent	Number Depts	Percent
1,000,000 or more	0	0.0%	0	0.0%	3	20.0%	10	80.0%	0	0.0%	13	100.0%
500,000 to 999,999	1	3.1	0	0.0	14	37.5	23	59.4	0	0.0	38	100.0
250,000 to 499,999	0	0.0	29	45.2	35	54.8	0	0.0	0	0.0	64	100.0
100,000 to 249,999	4	1.9	5	2.3	89	41.3	115	53.5	2	0.9	215	100.0
50,000 to 99,999	5	1.1	15	3.0	170	34.9	290	59.6	7	1.4	487	100.0
25,000 to 49,999	19	1.8	64	6.1	404	38.4	560	53.2	5	0.5	1,051	100.0
10,000 to 24,999	71	2.5	268	9.4	1,301	45.8	1,193	42.0	8	0.3	2,841	100.0
5,000 to 9,999	153	4.2	537	14.8	1,751	48.4	1,168	32.3	13	0.4	3,621	100.0
2,500 to 4,999	271	5.9	924	20.3	2,243	49.2	1,100	24.1	23	0.5	4,561	100.0
Under 2,500	1,615	12.1	3,733	28.0	5,585	41.8	2,360	17.7	55	0.4	13,350	100.0
Total	2,138	8.1	5,549	21.1	11,590	44.2	6,847	26.1	115	0.4	26,239	100.0

Source: FEMA US Fire Administration 2002
Survey of the Needs of the US Fire Service

The above projections are based on 8,364 departments reporting on Question 30b. Numbers may not add to totals due to rounding.

Q. 30b: How much of your personal protective clothing is at least 10 years old?

Table 41
Does Department Have Reserve Protective Clothing
Sufficient to Equip 10% of Emergency Responders?
by Community Size
(Q. 30c)

Population of Community	Yes Number Depts	Percent	No Number Depts	Percent	Don't Know Number Depts	Percent	Total Number Depts	Percent
1,000,000 or more	10	80.0%	3	20.0%	0	0.0%	13	100.0%
500,000 to 999,999	20	51.5	18	48.5	0	0.0	38	100.0
250,000 to 499,999	43	67.5	19	30.0	2	2.5	64	100.0
100,000 to 249,999	143	66.5	70	32.4	2	1.2	215	100.0
50,000 to 99,999	300	61.5	180	37.1	7	1.4	487	100.0
25,000 to 49,999	598	57.2	424	40.5	24	2.3	1,051	100.0
10,000 to 24,999	1,392	49.0	1,406	49.5	43	1.5	2,841	100.0
5,000 to 9,999	1,412	39.0	2,169	59.9	40	1.1	3,621	100.0
2,500 to 4,999	1,600	35.1	2,901	63.6	59	1.3	4,561	100.0
Under 2,500	4,272	32.0	8,730	65.4	347	2.6	13,350	100.0
Total	9,760	37.2	15,953	60.8	525	2.0	26,239	100.0

Source: FEMA US Fire Administration 2002
        Survey of the Needs of the US Fire Service

The above projections are based on 8,355 departments reporting on Question 30c. Numbers may not add to totals due to rounding.

Q. 30c: Do you have reserve personal protective clothing sufficient to equip 10% of your emergency responders?

COMMUNICATIONS AND COMMUNICATIONS EQUIPMENT

Table 42 (p. 87) indicates what fraction of departments can communicate by radio at incident scenes with their Federal, state or local partners. Interestingly, the percentage who can declines as the size of the department increases, the reverse of every other question so far. Specifically, only three-fifths of departments serving communities of 1,000,000 or more population can communicate with partners. This rises to two-thirds for communities with 100,000 to 499,999 population, to three-fourths for communities with 5,000 to 99,999 population, and to four-fifths for communities with less than 5,000 population.

Table 43 (p. 88) indicates what fraction of partners departments can communicate with, for those departments that indicated in the previous question that they can communicate with partners. Responses were similar across all community sizes, with two-fifths saying they can communicate with all partners, nearly half saying they can communicate with most partners, and one-sixth saying they can communicate only with some partners.

Tables 44 and 45 (pp. 89-90) collectively address the ability of fire departments to access a map coordinate system with sufficient standardization of format to provide effective functionality in directing the movements of emergency response partners.

Table 44 indicates that nearly half of all fire departments have no map coordinate system. This is a problem particularly for smaller communities, up to 99,999 population. About one-seventh of communities with at least 500,000 population have no map coordinate system.

Table 45 indicates that the vast majority of departments with a map coordinate system have only a local system, which means the system they have is unlikely to be usable with global positioning systems (GPS) or familiar to, or easily used by, non-local emergency response partners, such as Urban Search and Rescue Teams, the National Guard, and state or national response forces. Moreover, interoperability of spatial-based information systems, equipment, and procedures will likely be rendered impossible beyond the local community under these circumstances. This reliance almost exclusively on local systems exists across-the-board, in all sizes of communities.

Table 46 (p. 91) indicates the use of 911 type telephone communication systems. One-quarter (25%) of departments have 911 Basic (32% of rural fire departments, protecting less than 2,500 population). About two-thirds (69%) have 911 Enhanced (59% of rural fire departments). Less than 1% have some other, unspecified 3-digit system. And 6% have no system with a special 3-digit number (8% of rural fire departments).

Table 47 (p. 92) indicates who has primary responsibility for dispatch operations. Overall, only one department in 12 (8.7%) has that responsibility lodged with the fire department, but the percentage goes up sharply with the size of the community. It is only

6-8% for communities with less than 10,000 population but rises to 80% for communities of at least 1 million population.

Roughly one-third (32.5%) have responsibility lodged with the police department, and that is fairly consistent until the size of community reaches 250,000 or more. Overall, 1.8% have responsibility lodged with a private company.

Another one-third (34.3%) have responsibility lodged with a combined public safety agency, and that is fairly consistent until the size of community reaches 500,000 or more, where the percentage drops. And the other one-fifth to one-fourth (22.7%) have some other arrangement.

Table 48 (p. 93) indicates whether there is a backup dispatch facility. Two of every five departments (40%) say no. The percentage without such a facility is 47% for departments protecting less than 2,500 population, 35% for departments protecting 2,500 to 4,999 population, 28-31% for departments protecting 5,000 to 99,999 population, and 23-27% for the larger communities, except for those protecting 250,000 to 499,999 population, where the percentage without was 33%.

Table 49 (p. 94) indicates whether there is Internet access for the department. The overall percentage of departments with access is 58%, but the percentages are much higher for departments protecting 5,000 or more population.

Table 50 (p. 95) indicates what kind of access departments have. Roughly half (49.8%) of departments with access have it at the department's only fire station. Another 11.4% have individual access for all personnel. Another 10.3% have access at each of their several fire stations, and 15.1% have access only at headquarters despite having multiple fire stations. The rest (13.5%) have some other arrangement.

Table 42
Can Department Communicate by Radio at an Incident Scene
With Federal, State or Local Partners?
by Community Size
(Q. 31a)

	Yes		No		Don't Know		Total	
Population of Community	Number of Depts	Percent	Number Depts	Percent	Number Depts	Percent	Number Depts	Percent
1,000,000 or more	8	61.5%	5	38.5%	0	0.0%	13	100.0%
500,000 to 999,999	24	63.1	12	31.6	2	5.3	38	100.0
250,000 to 499,999	44	68.8	20	31.2	0	0.0	64	100.0
100,000 to 249,999	142	66.0	63	29.3	10	4.7	215	100.0
50,000 to 99,999	362	74.4	107	22.0	18	3.6	487	100.0
25,000 to 49,999	741	70.4	279	26.5	32	3.1	1,053	100.0
10,000 to 24,999	2,060	72.5	708	24.9	75	2.6	2,843	100.0
5,000 to 9,999	2,734	75.3	792	21.8	103	2.8	3,629	100.0
2,500 to 4,999	3,617	79.1	803	17.6	152	3.3	4,572	100.0
Under 2,500	11,034	82.1	1,881	14.0	523	3.9	13,440	100.0
Total	20,768	78.9	4,670	17.7	915	3.5	26,354	100.0

Source: FEMA US Fire Administration 2002
         Survey of the Needs of the US Fire Service

The above projections are based on 8,370 departments reporting on Question 31a. Numbers may not add to totals due to rounding.

Q. 31a: Can you communicate by radio on an incident scene with your federal, state and local emergency response partners (includes frequency compatability)?

Table 43
For Departments That Can Communicate With Partners at an Incident Scene
What Fraction of Partners Can They Communicate With?
by Community Size
(Q. 31b)

Population	All		Most		Some		Total	
of Community	Number Depts	Percent	Number Depts	Percent	Number Depts	Percent	Number Depts	Percent
1,000,000 or more	3	37.5%	0	0.0%	5	62.5%	8	100.0%
500,000 to 999,999	10	41.7	10	41.7	4	16.6	24	100.0
250,000 to 499,999	14	31.8	22	50.0	8	18.1	44	100.0
100,000 to 249,999	57	40.1	63	44.3	22	15.5	142	100.0
50,000 to 99,999	133	36.7	152	42.0	77	21.3	362	100.0
25,000 to 49,999	274	37.0	320	43.2	147	19.9	741	100.0
10,000 to 24,999	729	35.4	987	47.9	344	16.7	2,060	100.0
5,000 to 9,999	1,080	39.5	1,241	45.4	413	15.1	2,734	100.0
2,500 to 4,999	1,367	37.8	1,649	45.6	600	16.6	3,617	100.0
Under 2,500	4,336	39.3	4,855	44.0	1,854	16.8	11,034	100.0
Total	7,992	39.8	9,283	46.3	3,467	17.3	20,786	100.0

Source: FEMA US Fire Administration 2002
    Survey of the Needs of the US Fire Service

The above projections are based on 6,405 departments reporting yes to Question 31a and also reporting on Question 31b. Numbers may not add to totals due to rounding.

Q. 31b: If [you can communicate by radio on an incident scene with your federal, state, and local emergency response partners], how many of your partners can you communicate with at an incident scene?

Table 44
Does Department Have a Map Coordinate System
to Help Direct Emergency Response Partners?
by Community Size
(Q. 32a)

Population of Community	Yes Number Depts	Yes Percent	No Number Depts	No Percent	Don't Know Number Depts	Don't Know Percent	Total Number Depts	Total Percent
1,000,000 or more	10	76.9%	3	23.1%	0	0.0%	13	100.0%
500,000 to 999,999	32	84.2	5	13.2	1	2.6	38	100.0
250,000 to 499,999	45	70.3	19	29.7	0	0.0	64	100.0
100,000 to 249,999	154	71.6	58	27.0	3	1.4	215	100.0
50,000 to 99,999	267	53.4	217	44.6	9	1.9	487	100.0
25,000 to 49,999	526	49.9	519	49.3	9	0.8	1,053	100.0
10,000 to 24,999	1,223	43.0	1,567	55.1	52	1.8	2,843	100.0
5,000 to 9,999	1,770	48.8	1,796	49.5	62	1.7	3,629	100.0
2,500 to 4,999	2,391	52.3	2,080	45.5	101	2.2	4,572	100.0
Under 2,500	7,734	57.5	5,383	40.1	322	2.4	13,440	100.0
Total	14,154	53.7	11,641	44.2	559	2.1	26,354	100.0

Source: FEMA US Fire Administration 2002
Survey of the Needs of the US Fire Service

The above projections are based on 8,344 departments reporting on Question 32a. Numbers may not add to totals due to rounding.

Q. 32a: Do you have a map coordinate system you would use to help direct your emergency response partners to specific locations?

Table 45
For Departments That Have a Map Coordinate System
What System Do They Use?
by Community Size
(Q. 32b)

Population of Community	Longitude/Latitude		Local		Military Grid		State Plane Coordinate		Other		Total	
	Number Depts	Percent	Number Depts	Percent	Number Depts	Percent	Number Depts	Percent	Number Depts	Percent	Number Depts	Percent
1,000,000 or more	1	10.0%	9	90.0%	0	0.0%	0	0.0%	0	0.0%	10	100.0%
500,000 to 999,999	3	9.4	29	90.6	0	0.0	0	0.0	0	0.0	32	100.0
250,000 to 499,999	6	13.3	39	86.7	0	0.0	0	0.0	0	0.0	45	100.0
100,000 to 249,999	7	4.5	134	87.1	0	0.0	9	5.8	4	2.6	154	100.0
50,000 to 99,999	30	11.2	224	84.0	3	1.0	5	1.9	5	1.9	267	100.0
25,000 to 49,999	45	8.6	446	84.8	2	0.3	9	1.7	24	4.6	526	100.0
10,000 to 24,999	106	8.7	1,064	86.9	10	0.8	13	1.1	31	2.5	1,223	100.0
5,000 to 9,999	214	12.1	1,473	83.2	12	0.7	12	0.7	57	3.2	1,770	100.0
2,500 to 4,999	237	9.9	2,020	84.5	29	1.2	31	1.3	74	3.1	2,391	100.0
Under 2,500	595	7.7	6,652	86.1	69	0.9	85	1.1	325	4.2	7,734	100.0
Total	1,251	8.8	12,084	85.4	124	0.8	162	1.1	520	3.7	14,154	100.0

Source: FEMA US Fire Administration 2002
Survey of the Needs of the US Fire Service

The above projections are based on 4,327 departments reporting yes to Questions 32a and also reporting on Question 32b. Numbers may not add to totals due to rounding.

Q. 32b: If [you have a map coordinate system you would use to help direct your emergency response partners to specific locations], what system do you use? "Local system" includes map grid, street address, and box alarm number.

Table 46
Does Department Have 911 or Similar System?
by Community Size
(Q. 33)

Population of Community	Yes – 911 Basic		Yes – 911 Enhanced		Yes – Other 3-Digit System		No		Total	
	Number Depts	Percent	Number Depts	Percent	Number Depts	Percent	Number Depts	Percent	Number Depts	Percent
1,000,000 or more	1	7.7%	12	92.3%	0	0.0%	0	0.0%	13	100.0%
500,000 to 999,999	2	5.3	36	94.7	0	0.0	0	0.0	38	100.0
250,000 to 499,999	6	9.3	59	90.7	0	0.0	0	0.0	64	100.0
100,000 to 249,999	14	6.5	199	92.6	1	0.5	1	0.5	215	100.0
50,000 to 99,999	25	5.1	462	94.9	0	0.0	0	0.0	487	100.0
25,000 to 49,999	106	10.1	937	88.9	3	0.3	7	0.6	1,053	100.0
10,000 to 24,999	420	14.8	2,378	83.6	6	0.2	40	1.4	2,843	100.0
5,000 to 9,999	700	19.3	2,799	77.1	5	0.1	125	3.4	3,629	100.0
2,500 to 4,999	1,076	23.5	3,274	71.6	6	0.1	215	4.7	4,572	100.0
Under 2,500	4,342	32.3	7,915	58.9	65	0.5	1,117	8.3	3,440	100.0
Total	6,697	25.4	18,064	68.6	87	0.3	1,504	5.7	6,354	100.0

Source: FEMA US Fire Administration 2002
   Survey of the Needs of the US Fire Service

The above projections are based on 8,358 reporting on Question 33. Numbers may not add to totals due to rounding.

Q. 33: Do you have 911 or similar system?

91

Table 47
Who Has Primary Responsibility for Dispatch Operations?
by Community Size
(Q. 34a)

Population of Community	Fire Department Number Depts	Percent	Police Department Number Depts	Percent	Private Company Number Depts	Percent	Combined Public Safety Agency Number Depts	Percent	Other Number Depts	Percent	Total Number Depts	Percent
1,000,000 or more	10	76.9%	2	15.4%	0	0.0%	0	0.0%	1	7.7%	13	100.0%
500,000 to 999,999	25	65.8	3	7.9	0	0.0	6	15.8	4	10.5	38	100.0
250,000 to 499,999	25	39.1	10	15.7	0	0.0	21	32.8	8	12.5	64	100.0
100,000 to 249,999	66	30.6	66	30.6	1	0.5	62	28.9	20	9.3	215	100.0
50,000 to 99,999	90	18.5	169	34.7	13	2.8	154	31.7	60	12.4	487	100.0
25,000 to 49,999	158	15.0	366	34.7	9	1.8	376	35.7	145	13.7	1,053	100.0
10,000 to 24,999	343	12.1	1,079	38.0	64	2.2	880	30.9	477	16.8	2,843	100.0
5,000 to 9,999	223	6.2	1,189	32.8	70	1.9	1,308	36.1	839	23.1	3,629	100.0
2,500 to 4,999	328	7.2	1,348	29.5	86	1.9	1,714	37.5	1,095	23.9	4,572	100.0
Under 2,500	1,022	7.6	4,342	32.3	228	1.7	4,523	33.7	3,324	24.7	13,440	100.0
Total	2,293	8.7	8,572	32.5	470	1.8	9,044	34.3	5,972	22.7	26,354	100.0

Source: FEMA US Fire Administration 2002
Survey of the Needs of the US Fire Service

The above table breakdown and projections are based on 8,334 reporting on Question 34a. Numbers may not add to totals due to rounding.

Q. 34a: Who has primary responsibility for dispatch operations?

Table 48
Does Department Have a Backup Dispatch Facility?
by Community Size
(Q. 34b)

Population	Yes		No		Total	
of Community	Number Depts	Percent	Number Depts	Percent	Number Depts	Percent
1,000,000 or more	10	76.9%	3	23.1%	13	100.0%
500,000 to 999,999	28	73.7	10	26.3	38	100.0
250,000 to 499,999	43	67.2	21	32.8	64	100.0
100,000 to 249,999	157	73.0	58	27.0	215	100.0
50,000 to 99,999	338	69.4	149	30.6	487	100.0
25,000 to 49,999	759	72.1	294	27.9	1,053	100.0
10,000 to 24,999	1,994	70.1	849	29.9	2,843	100.0
5,000 to 9,999	2,504	69.0	1,125	31.0	3,629	100.0
2,500 to 4,999	2,960	64.7	1,612	35.3	4,572	100.0
Under 2,500	7,145	53.2	6,295	46.8	13,440	100.0
Total	15,940	60.5	10,414	39.5	26,354	100.0

Source: FEMA US Fire Administration 2002
      Survey of the Needs of the US Fire Service

The above projections are based on 7,239 departments reporting on Question 34a. Numbers may not add to totals due to rounding.

Q. 34a: Who has primary responsibility for dispatch operations?

Table 49
Does Department Have Internet Access?
by Community Size
(Q. 35a)

Population	Yes		No		Total	
of Community	Number Depts	Percent	Number Depts	Percent	Number Depts	Percent
1,000,000 or more	13	100.0%	0	0.0%	13	100.0%
500,000 to 999,999	38	100.0	0	0.0	38	100.0
250,000 to 499,999	64	100.0	0	0.0	64	100.0
100,000 to 249,999	211	98.1	4	1.9	215	100.0
50,000 to 99,999	471	96.7	16	3.3	487	100.0
25,000 to 49,999	981	93.2	72	6.8	1,053	100.0
10,000 to 24,999	2,480	87.3	362	12.7	2,843	100.0
5,000 to 9,999	2,755	75.9	874	24.1	3,629	100.0
2,500 to 4,999	2,863	62.6	1,709	37.4	4,572	100.0
Under 2,500	5,472	40.7	7,968	59.3	13,440	100.0
Total	15,347	58.2	11,007	41.8	26,354	100.0

Source: FEMA US Fire Administration 2002
       Survey of the Needs of the US Fire Service

The above table breakdown and projections are based on 8,302 departments reporting on Question 35a. Numbers may not add to totals due to rounding.

Q. 35a: Does your department have Internet access?

Table 50
For Departments That Have Internet Access
What Kind of Access Do They Have?
by Community Size
(Q. 35b)

Population of Community	All Personnel Have Individual Access		One Access Point per Station – Multiple Stations		One Access Point at the Only Station		Access at Headquarters – Multiple Stations		Other		Total	
	Number Depts	Percent	Number Depts	Percent	Number Depts	Percent	Number Depts	Percent	Number Depts	Percent	Number Depts	Percent
1,000,000 or more	1	7.7%	4	30.8%	0	0.0%	4	30.8%	4	30.8%	13	100.0%
500,000 to 999,999	8	21.0	17	44.7	0	0.0	9	23.7	4	10.5	38	100.0
250,000 to 499,999	12	18.8	23	35.9	0	0.0	21	32.8	8	12.5	64	100.0
100,000 to 249,999	63	29.9	86	40.8	0	0.0	52	24.5	11	5.3	211	100.0
50,000 to 99,999	132	28.0	163	34.6	0	0.0	155	32.8	21	4.6	471	100.0
25,000 to 49,999	253	25.8	256	26.1	102	10.4	293	29.9	77	7.8	981	100.0
10,000 to 24,999	389	15.7	459	18.5	806	32.5	692	27.9	134	5.4	2,480	100.0
5,000 to 9,999	300	10.9	237	8.6	1,623	58.9	424	15.4	174	6.3	2,755	100.0
2,500 to 4,999	195	6.8	146	5.1	1,918	67.0	298	10.4	309	10.8	2,863	100.0
Under 2,500	389	7.1	180	3.3	3,195	58.4	361	6.6	1,340	24.5	5,472	100.0
Total	1,734	11.4	1,562	10.3	7,590	49.8	2,299	15.1	2,060	13.5	15,347	100.0

Source: FEMA US Fire Administration 2002
Survey of the Needs of the US Fire Service

The above projections are based on 5,482 departments reporting yes to Question 35a and also reporting on Question 35b. Numbers may not add to totals due to rounding.

Q. 35b: If [your department has Internet access], describe the access you have.

ABILITY TO HANDLE UNUSUALLY CHALLENGING INCIDENTS

Questions 36-39 were designed to check the capabilities of fire departments, in communities of various sizes, to handle unusually severe and challenging incidents, only one of which involved a fire. These have to do with the increasingly important first responder role of fire departments.

In addition to asking whether such incidents were within the department's scope, the survey asked whether fire departments could handle such incidents with local personnel and equipment and whether a plan existed to support effective coordination with non-local resources and partners.

Technical Rescue and EMS at Structural Collapse With 50 Occupants

Table 51 (p. 110) indicates whether a technical rescue with EMS at a structural collapse of a building with 50 occupants is within the scope of the department.

Affirmative answers become less likely as the size of the department shrinks, so that less than half the fire departments protecting rural communities (less than 2,500 population) answered affirmatively, while all of the largest departments (those protecting at least 500,000 population) did.

Tables 52-54 (pp. 111-113) address, for the departments that consider such a rescue within their scope, how far they have to go for people and equipment and whether they have a plan, respectively.

By combining Table 51 with Tables 52-54, one can obtain an even better indication of different types of department needs to address such incidents, as seen in Tables S to U. In Tables S to U, the rightmost column reproduces the "No, not within scope" statistics from Table 51. The other columns are produced by multiplying the columns from Tables 52-54, respectively, by the "Yes, within scope" statistics from Table 51.

For communities with at least 5,000 but less than 500,000 population, the majority of fire departments will look elsewhere for at least some of the specially trained people they would need.

In the smaller communities, most departments that consider such an incident within their scope would need non-local personnel, but they do not constitute a majority of departments because so many departments do not consider such incidents within their scope. For communities of 10,000 to 24,999 population, 27% of departments do not consider an incident such as this to be within their scope. For communities of 5,000 to 9,999 population, the percentage of departments considering such an incident outside their scope rises to 32%, and to 39% for communities of 2,500 to 4,999 population. For communities with less than 2,500 population, the majority of departments (56%) consider such an incident outside their scope.

97

Table S. Departments by Whether They Can Handle This Type of Incident, Where They Obtain Necessary Personnel With Specialized Training, and Size of Community (Q. 36b)

Population Protected	Can Department Handle Technical Rescue with EMS at Structural Collapse of a Building with 50 Occupants?		
	Yes and With Local Trained People	Yes But Need Non-Local Trained People	No, Not Within Scope
1,000,000 or more	91%	9%	0%
500,000 to 999,999	63%	37%	0%
250,000 to 499,999	42%	46%	12%
100,000 to 249,999	25%	61%	14%
50,000 to 99,999	17%	66%	17%
25,000 to 49,999	12%	70%	18%
10,000 to 24,999	11%	62%	27%
5,000 to 9,999	13%	55%	32%
2,500 to 4,999	13%	48%	39%
Under 2,500	10%	35%	56%
Total	11%	45%	44%

The above projections are based on 8,268 departments reporting on Question 36a and 5,146 reporting on Question 36b. See Tables 51 and 52.

Q. 36b: If [technical rescue and EMS for a building with 50 occupants after structural collapse is within your department's scope], how far would you have to go to obtain enough people with specialized training for this incident?

Except for communities with 500,000 or more population, at least half of fire departments must look elsewhere for at least some of the specialized equipment they would need – except for communities with less than 2,500 population, where the majority of departments do not consider an incident like this within their scope. For communities with less than 2,500 population, the majority of departments (56%) consider such an incident outside their scope.

For communities with 1 million or more population, 77% of departments will handle such an incident and require only locally available equipment. The percentage is 63% for communities with 500,000 to 999,999 population, 37% for communities with 250,000 to 499,999 population, and 25% for communities with 100,000 to 249,999 population.

For communities with 10,000 to 99,999 population, 63-70% of fire departments will look elsewhere for specialized equipment, and 17-27% will not be looking because such an incident is not within their scope. Only 10-17% consider such an incident within their scope and are prepared to address it with only locally available equipment.

Table T. Departments by Whether They Can Handle This Type of Incident, Where They Obtain the Necessary Specialized Equipment, and Size of Community (Q. 36c)

Population Protected	Can Department Handle Technical Rescue with EMS at Structural Collapse of a Building with 50 Occupants?		
	Yes and With Local Equipment	Yes But Need Non-Local Equipment	No, Not Within Scope
1,000,000 or more	77%	23%	0%
500,000 to 999,999	63%	37%	0%
250,000 to 499,999	37%	51%	12%
100,000 to 249,999	25%	61%	14%
50,000 to 99,999	17%	66%	17%
25,000 to 49,999	12%	70%	18%
10,000 to 24,999	10%	63%	27%
5,000 to 9,999	12%	56%	32%
2,500 to 4,999	11%	50%	39%
Under 2,500	9%	35%	56%
Total	11%	46%	44%

The above table breakdown and projections are based on 8,268 departments reporting on Question 36a and 5,083 reporting on Question 36c. See Tables 51and 53.

Q. 36c: If [technical rescue and EMS for a building with 50 occupants after structural collapse is within your department's scope], how far would you have to go to obtain enough specialized equipment to handle this incident?

At least 64% of departments protecting communities of less than 50,000 population either would not consider such an incident within their scope or do not have a written agreement in place to obtain and use resources in a timely, efficient and effective manner for such an incident.

Many departments that consider such an incident within their scope have no agreement of any kind.

Written agreements are in place for 92% of communities with at least 1 million population, for 58% of communities with 500,000 to 999,999 population, for 67% of communities with 250,000 to 499,999 population, for roughly half (47-50%) of communities with 50,000 to 249,999 population.

Written agreements are in place for only one-third (30-36%) of communities with 10,000 to 49,999 population, only one-fourth (23%) of communities with 5,000 to 9,999 population, only one-fifth (18%) of communities with 2,500 to 4,999 population, and only one-twelfth (13%) of communities with less than 2,500 population.

Table U. Departments by Whether They Can Handle This Type of Incident, Type of Plan for Using Non-Local Resources, and Size of Community (Q. 36d)

Population Protected	Can Department Handle Technical Rescue with EMS at Structural Collapse of a Building with 50 Occupants?			
	Yes – Written Agreement	Yes – But Not Written	Yes – But No Plan	No, Not Within Scope
1,000,000 or more	92%	8%	0%	0%
500,000 to 999,999	58%	42%	0%	0%
250,000 to 499,999	67%	22%	0%	12%
100,000 to 249,999	50%	30%	6%	14%
50,000 to 99,999	47%	29%	8%	17%
25,000 to 49,999	36%	33%	13%	18%
10,000 to 24,999	30%	32%	12%	27%
5,000 to 9,999	23%	33%	12%	32%
2,500 to 4,999	18%	30%	14%	39%
Under 2,500	13%	21%	11%	56%
Total	19%	26%	11%	44%

The above table breakdown and projections are based on 8,268 departments reporting on Question 36a and 5,090 reporting on Question 36d. See Tables 51 and 54.

Q. 36d: If [technical rescue and EMS for a building with 50 occupants after structural collapse is within your department's scope], do you have a plan for working with others on this type of incident?

Hazmat and EMS for Incident Involving Chemical/Biological Agents and 10 Injuries

Table 55 indicates whether hazmat and EMS for an incident involving chemical/ biological agents and 10 injuries is within the scope of the department. (Note that casualty counts of 100 to 1,000 are not unusual in chemical/biological agent weapons of mass destruction.)

Affirmative answers become less likely as the size of the department shrinks, so that less than half the fire departments protecting rural communities (less than 2,500 population) answered affirmatively, while all of the largest departments (those protecting at least 500,000 population) did.

Tables 56-58 address, for the departments that consider such an incident within their scope, how far they have to go for people and equipment and whether they have a plan, respectively.

By combining Table 55 with Tables 56-58, one can obtain an even better indication of different types of department needs to address such incidents, as seen in Tables V to X.

In Tables V to X, the rightmost column reproduces the "No, not within scope" statistics from Table 55. The other columns are produced by multiplying the columns from Tables 56-58, respectively, by the "Yes, within scope" statistics from Table 55.

Table V. Departments by Whether They Can Handle This Type of Incident, Where They Obtain Necessary Personnel With Specialized Training, and Size of Community (Q. 37b)

Population Protected	Can Department Handle a Hazmat and EMS Incident Involving Chemical/Biological Agents and 10 Injuries?		
	Yes and With Local Trained People	Yes But Need Non-Local Trained People	No, Not Within Scope
1,000,000 or more	100%	0%	0%
500,000 to 999,999	90%	11%	0%
250,000 to 499,999	73%	25%	2%
100,000 to 249,999	56%	37%	7%
50,000 to 99,999	33%	59%	8%
25,000 to 49,999	27%	58%	16%
10,000 to 24,999	18%	59%	24%
5,000 to 9,999	14%	54%	32%
2,500 to 4,999	11%	49%	40%
Under 2,500	9%	37%	53%
Total	13%	45%	42%

The above table breakdown and projections are based on 8,250 departments reporting on Question 37a and 5,292 reporting on Question 37b. See Tables 55 and 56.

Q. 37b: If [hazmat and EMS for an incident involving chemical/biological agents and 10 injuries is within your department's scope], how far would you have to go to obtain enough people with specialized training for this incident?

Except for communities with 100,000 or more population, the majority of fire departments will look elsewhere for at least some of the specially trained people they would need – except for communities with less than 5,000 population, where at least 40% of departments do not consider an incident like this within their scope.

One-sixth of communities with 25,000 to 49,999 population (16%) consider an incident such as this outside their scope. One-fourth (24%) of communities with 10,000 to 24,999 population consider such an incident outside their scope. The percentage of communities where departments consider such incidents outside their scope rises to one-third (32%) for communities with 5,000 to 9,999 population and to roughly half (40-53%) for communities with less than 5,000 population.

Table W. Departments by Whether They Can Handle This Type of Incident, Where They Obtain the Necessary Specialized Equipment, and Size of Community (Q. 37c)

Population Protected	Can Department Handle a Hazmat and EMS Incident Involving Chemical/Biological Agents and 10 Injuries?		
	Yes and With Local Equipment	Yes But Need Non-Local Equipment	No, Not Within Scope
1,000,000 or more	92%	8%	0%
500,000 to 999,999	82%	18%	0%
250,000 to 499,999	69%	30%	2%
100,000 to 249,999	51%	42%	7%
50,000 to 99,999	29%	63%	8%
25,000 to 49,999	21%	63%	16%
10,000 to 24,999	16%	61%	24%
5,000 to 9,999	12%	57%	32%
2,500 to 4,999	9%	51%	40%
Under 2,500	8%	39%	53%
Total	11%	47%	42%

The above projections are based on 8,250 departments reporting on Question 37a and 5,239 reporting on Question 37c. See Tables 55 and 57.

Q. 37c: If [hazmat and EMS for an incident involving chemical/biological agents and 10 injuries is within your department's scope], how far would you have to go to obtain enough specialized equipment to handle this incident?

Except for communities with 100,000 or more population, the majority of fire departments must look elsewhere for at least some of the specialized equipment they would need – except for communities with less than 2,500 population, where the majority of departments do not consider an incident like this within their scope.

For communities with 10,000 to 99,999 population, 61-63% of fire departments will look elsewhere for specialized equipment, and many of the rest will not be looking only because such an incident is not within their scope.

At least one-third of departments protecting communities of less than 500,000 population either would not consider such an incident within their scope or do not have a written agreement in place to obtain and use resources in a timely, efficient and effective manner for such an incident.

Many departments that consider such an incident within their scope have no agreement of any kind.

Table X. Departments by Whether They Can Handle This Type of Incident, Type of Plan for Using Non-Local Resources, and Size of Community (Q. 37d)

Population Protected	Can Department Handle a Hazmat and EMS Incident Involving Chemical/Biological Agents and 10 Injuries?			
	Yes – Written Agreement	Yes – But Not Written	Yes – But No Plan	No, Not Within Scope
1,000,000 or more	92%	8%	0%	0%
500,000 to 999,999	71%	29%	0%	0%
250,000 to 499,999	67%	31%	0%	2%
100,000 to 249,999	64%	27%	2%	7%
50,000 to 99,999	62%	25%	5%	8%
25,000 to 49,999	51%	27%	6%	16%
10,000 to 24,999	36%	34%	6%	24%
5,000 to 9,999	25%	34%	9%	32%
2,500 to 4,999	18%	32%	11%	40%
Under 2,500	13%	24%	10%	53%
Total	21%	28%	9%	42%

The above projections are based on 8,250 departments reporting on Question 37a and 5,227 reporting on Question 37d. See Tables 55 and 58.

Q. 37d: If [hazmat and EMS for an incident involving chemical/biological agents and 10 injuries is within your department's scope], do you have a plan for working with others on this type of incident?

Wildland/Urban Interface Fire Affecting 500 Acres

Table 59 indicates whether a wildland/urban interface fire affecting 500 acres is within the scope of the department.

Affirmative answers are more likely for both the larger communities (at least 500,000 population) and the most rural communities (less than 10,000 population). However, a majority of all sizes of communities say such incidents are within their scope.

Tables 60-62 address, for the departments that consider such an incident within their scope, how far they have to go for people and equipment and whether they have a plan, respectively.

By combining Table 59 with Tables 60-62, one can obtain an even better indication of different types of department needs to address such incidents, as seen in Tables Y to AA.

In Tables Y to AA, the rightmost column reproduces the "No, not within scope" statistics from Table 59. The other columns are produced by multiplying the columns from Tables 60-62, respectively, by the "Yes, within scope" statistics from Table 59.

Table Y. Departments by Whether They Can Handle This Type of Incident, Where They Obtain Necessary Personnel With Specialized Training, and Size of Community (Q. 38b)

Population Protected	Can the Department Handle a Wildland/Urban Interface Fire Affecting 500 Acres?		
	Yes and With Local Trained People	Yes But Need Non-Local Trained People	No, Not Within Scope
1,000,000 or more	31%	38%	31%
500,000 to 999,999	37%	42%	21%
250,000 to 499,999	25%	36%	39%
100,000 to 249,999	18%	43%	39%
50,000 to 99,999	16%	41%	43%
25,000 to 49,999	14%	38%	48%
10,000 to 24,999	17%	44%	39%
5,000 to 9,999	21%	47%	31%
2,500 to 4,999	24%	48%	28%
Under 2,500	30%	42%	28%
Total	26%	44%	31%

The above projections are based on 8,248 departments reporting on Question 38a and 5,503 reporting on Question 38b. See Tables 59 and 60.

Q. 38b: If [wildland/urban interface fire affecting 500 acres is within your department's scope], how far would you have to go to obtain enough people with specialized training for this incident?

Wildland/urban interface fires have a much different profile from the first two types of major incidents.

A large share of departments in every community-population interval do not consider such incidents within their scope, and the majority who do in every interval do not consider their local personnel sufficient.

Table Z. Departments by Whether They Can Handle This Type of Incident, Where They Obtain the Necessary Specialized Equipment, and Size of Community (Q. 38c)

Population Protected	Can the Department Handle a Wildland/Urban Interface Fire Affecting 500 Acres?		
	Yes and With Local Equipment	Yes But Need Non-Local Equipment	No, Not Within Scope
1,000,000 or more	31%	38%	31%
500,000 to 999,999	26%	53%	21%
250,000 to 499,999	23%	37%	39%
100,000 to 249,999	15%	46%	39%
50,000 to 99,999	15%	43%	43%
25,000 to 49,999	12%	40%	48%
10,000 to 24,999	16%	45%	39%
5,000 to 9,999	18%	51%	31%
2,500 to 4,999	22%	50%	28%
Under 2,500	26%	46%	28%
Total	22%	47%	31%

The above projections are based on 8,248 departments reporting on Question 38a and 5,468 reporting on Question 38c. See Tables 59 and 61.

Q. 38c: If [wildland/urban interface fire affecting 500 acres is within your department's scope], how far would you have to go to obtain enough specialized equipment to handle this incident?

An even more dominant majority of departments considering such incidents within their scope in each population interval do not consider their local specialized equipment sufficient to handle such an incident.

This widely shared lack of local sufficiency has been recognized for many years by the US Forest Service, which has applied considerable effort to create formal networks and plans to move resources from wherever they are to wherever they are needed. Their efforts appear to have borne fruit, because the current state of planning is also much different for these incidents – and specifically much better.

For communities with less than 10,000 population where the department considers such an incident within its scope, nearly half of the departments have written plans for accessing and using non-local personnel and equipment.

For communities with at least 10,000 population where the department considers such an incident within its scope, considerably more than half the departments have written plans.

Table AA. Departments by Whether They Can Handle This Type of Incident, Type of Plan for Using Non-Local Resources, and Size of Community (Q. 38d)

Population Protected	Can the Department Handle a Wildland/Urban Interface Fire Affecting 500 Acres?			
	Yes – Written Agreement	Yes – But Not Written	Yes – But No Plan	No, Not Within Scope
1,000,000 or more	69%	0%	0%	31%
500,000 to 999,999	50%	21%	8%	21%
250,000 to 499,999	42%	16%	3%	39%
100,000 to 249,999	43%	15%	2%	39%
50,000 to 99,999	40%	13%	4%	43%
25,000 to 49,999	34%	15%	3%	48%
10,000 to 24,999	33%	23%	5%	39%
5,000 to 9,999	33%	30%	5%	31%
2,500 to 4,999	32%	34%	6%	28%
Under 2,500	32%	35%	5%	28%
Total	33%	31%	5%	31%

The above projections are based on 8,248 departments reporting on Question 38a and 5,448 reporting on Question 38d. See Tables 59 and 62.

Q. 38d: If [wildland/urban interface fire affecting 500 acres is within your department's scope], do you have a plan for working with others on this type of incident?

Mitigation of a Developing Major Flood

Table 63 indicates whether mitigation of a developing major flood is within the scope of the department.

Affirmative answers become less likely as the size of the department shrinks, so that less than half the fire departments protecting communities with less than 5,000 population answered affirmatively, while two-thirds of the departments protecting communities with at least 500,000 population did.

Tables 64-66 address, for the departments that consider such an incident within their scope, how far they have to go for people and equipment and whether they have a plan, respectively.

By combining Table 63 with Tables 64-66, one can obtain an even better indication of different types of department needs to address such incidents, as seen in Tables AB to AD.

In Tables AB to AD, the rightmost column reproduces the "No, not within scope" statistics from Table 63. The other columns are produced by multiplying the columns from Tables 64-66, respectively, by the "Yes, within scope" statistics from Table 63.

Table AB. Departments by Whether They Can Handle This Type of Incident, Where They Obtain Necessary Personnel With Specialized Training, and Size of Community (Q. 39b)

	Can the Department Handle Mitigation of a Developing Major Flood?		
Population Protected	Yes and With Local Trained People	Yes But Need Non-Local Trained People	No, Not Within Scope
1,000,000 or more	54%	15%	31%
500,000 to 999,999	39%	29%	32%
250,000 to 499,999	19%	47%	34%
100,000 to 249,999	14%	50%	36%
50,000 to 99,999	11%	51%	38%
25,000 to 49,999	12%	46%	42%
10,000 to 24,999	11%	43%	46%
5,000 to 9,999	12%	41%	47%
2,500 to 4,999	14%	35%	51%
Under 2,500	12%	27%	61%
Total	12%	33%	54%

The above projections are based on 8,162 departments reporting on Question 39a and 3,967 reporting on Question 39b. See Tables 63 and 64.

Q. 39b: If [mitigation (confining, slowing, etc.) of a developing major flood is within your department's scope], how far would you have to go to obtain enough people with specialized training for this incident?

Except for communities with 500,000 or more population, less than one-fifth of fire departments are prepared to handle mitigation of a developing major flood with local specialized people.

Far more departments (two to five times as many) consider such an incident within their scope but would have to use non-local people.

Significant fractions of departments in all community-size groups consider such an incident to be outside their scope.

This is true for one-third (31-34%) of communities with at least 250,000 population and slightly more than one-third (36-38%) of communities with 50,000 to 249,999 population.

Departments considering such an incident to be outside their scope account for nearly half (42-47%) of communities with 5,000 to 49,999 population, half (51%) of communities with 2,500 to 4,999 population, and three-fifths (61%) of communities with less than 2,500 population.

Table AC. Departments by Whether They Can Handle This Type of Incident, Where They Obtain the Necessary Specialized Equipment, and Size of Community (Q. 39c)

Population Protected	Can the Department Handle Mitigation of a Developing Major Flood?		
	Yes and With Local Equipment	Yes But Need Non-Local Equipment	No, Not Within Scope
1,000,000 or more	23%	46%	31%
500,000 to 999,999	32%	37%	32%
250,000 to 499,999	17%	48%	34%
100,000 to 249,999	11%	53%	36%
50,000 to 99,999	9%	53%	38%
25,000 to 49,999	10%	49%	42%
10,000 to 24,999	10%	44%	46%
5,000 to 9,999	10%	43%	47%
2,500 to 4,999	12%	37%	51%
Under 2,500	10%	28%	61%
Total	11%	35%	54%

The above projections are based on 8,162 departments reporting on Question 39a and 3,947 reporting on Question 39c. See Tables 63 and 65.

Q. 39c: If [mitigation (confining, slowing, etc.) of a developing major flood is within your department's scope], how far would you have to go to obtain enough specialized equipment to handle this incident?

Except for communities with 500,000 or more population, less than one-fifth of fire departments are prepared to handle mitigation of a developing major flood with local equipment.

Far more departments (three to five times as many) consider such an incident within their scope but would have to use non-local equipment.

108

Except for communities with 100,000 to 499,999 population, there are more departments that consider such an incident within their scope but have no written plan than there are departments that have a written plan.

For communities with at least 50,000 population, only 30-39% of departments have a written plan.

The percentages are lower for smaller communities – 25% for communities of 25,000 to 49,999 population, 20% for communities of 10,000 to 24,999 population, 16% for communities of 5,000 to 9,999 population, 13% for communities of 2,500 to 4,999 population, and only 8% for communities with less than 2,500 population.

Combined with the results on local resources, this suggests that, of the four reference major incidents included on the Needs Assessment Survey, mitigation of a major flood is the one that most combines an insufficiency of local resources with a lack of written plans to effectively access and use non-local resources, across the entire spectrum of community sizes.

Table AD. Departments by Whether They Can Handle This Type of Incident, Type of Plan for Using Non-Local Resources, and Size of Community (Q. 39d)

Population Protected	Can the Department Handle Mitigation of a Developing Major Flood?			
	Yes – Written Agreement	Yes – But Not Written	Yes – But No Plan	No, Not Within Scope
1,000,000 or more	31%	31%	8%	31%
500,000 to 999,999	32%	37%	0%	32%
250,000 to 499,999	39%	25%	2%	34%
100,000 to 249,999	36%	22%	6%	36%
50,000 to 99,999	36%	18%	8%	38%
25,000 to 49,999	25%	21%	13%	42%
10,000 to 24,999	20%	23%	12%	46%
5,000 to 9,999	16%	24%	13%	47%
2,500 to 4,999	13%	24%	13%	51%
Under 2,500	8%	20%	11%	61%
Total	13%	21%	11%	54%

The above projections are based on 8,162 departments reporting on Question 39a and 3,916 reporting on Question 39d. See Tables 63 and 66.

Q. 39d: If [mitigation (confining, slowing, etc.) of a developing major flood is within your department's scope], do you have a plan for working with others on this type of incident?

Table 51
Is Technical Rescue and EMS for a Building
With 50 Occupants After Structural Collapse
Within the Scope of Department?
by Community Size
(Q. 36a)

Population of Community	Yes Number Depts	Yes Percent	No Number Depts	No Percent	Total Number Depts	Total Percent
1,000,000 or more	13	100.0%	0	0.0%	13	100.0%
500,000 to 999,999	38	100.0	0	0.0	38	100.0
250,000 to 499,999	57	88.4	7	11.6	64	100.0
100,000 to 249,999	184	85.7	31	14.3	215	100.0
50,000 to 99,999	405	83.2	82	16.9	487	100.0
25,000 to 49,999	860	81.6	193	18.4	1,053	100.0
10,000 to 24,999	2,090	73.5	753	26.5	2,843	100.0
5,000 to 9,999	2,475	68.2	1,154	31.8	3,629	100.0
2,500 to 4,999	2,787	61.0	1,785	39.0	4,572	100.0
Under 2,500	5,961	44.4	7,478	55.6	13,440	100.0
Total	14,872	56.4	11,482	43.6	26,354	100.0

Source: FEMA US Fire Administration 2002
       Survey of the Needs of the US Fire Service

The above projections are based on 8,268 departments reporting on Question 36a.
Numbers may not add to totals due to rounding.

Q. 36a: Is [technical rescue and EMS for a building with 50 occupants after structural collapse] within your department's scope?

Table 52
For Departments Where Technical Rescue and EMS For a Building With 50 Occupants After Structural Collapse Is Within Their Scope, How Far Do They Have to Go to Obtain Sufficient People With Specialized Training to Handle Such an Incident? by Community Size (Q. 36b)

Population of Community	Local		Regional		State		National		Total	
	Number Depts	Percent	Number Depts	Percent	Number Depts	Percent	Number Depts	Percent	Number Depts	Percent
1,000,000 or more	12	92.3%	0	0.0%	1	7.7%	0	0.0%	13	100.0%
500,000 to 999,999	24	63.2	8	21.0	4	10.5	2	5.2	38	100.0
250,000 to 499,999	27	47.4	24	42.1	4	7.0	2	3.5	57	100.0
100,000 to 249,999	54	29.3	95	51.6	29	15.8	6	3.3	184	100.0
50,000 to 99,999	85	20.9	238	58.8	70	17.3	12	3.0	405	100.0
25,000 to 49,999	127	14.8	505	58.7	185	21.5	43	5.0	860	100.0
10,000 to 24,999	315	15.1	1,213	57.9	506	24.3	56	2.7	2,090	100.0
5,000 to 9,999	463	18.8	1,387	56.0	580	23.4	44	1.8	2,475	100.0
2,500 to 4,999	570	20.5	1,603	57.5	576	20.7	38	1.4	2,787	100.0
Under 2,500	1,324	22.2	3,399	57.0	1,176	19.7	61	1.0	5,961	100.0
Total	3,002	20.2	8,470	57.0	3,134	21.1	264	1.8	4,872	100.0

Source: FEMA US Fire Administration 2002
Survey of the Needs of the US Fire Service

The above projections are based on 5,146 departments reporting yes to Question 36a and also reporting on Question 36b. Numbers may not add to totals due to rounding.

Q. 36b: If [technical rescue and EMS for a building with 50 occupants after structural collapse is within your department's scope], how far would you have to go to obtain enough people with specialized training for this incident?

111

Table 53
For Departments Where Technical Rescue and EMS For a Building With 50 Occupants After Structural Collapse Is Within Their Scope, How Far Do They Have to Go to Obtain Sufficient Specialized Equipment to Handle Such an Incident?
by Community Size
(Q. 36c)

Population of Community	Local Number Depts	Local Percent	Regional Number Depts	Regional Percent	State Number Depts	State Percent	National Number Depts	National Percent	Total Number Depts	Total Percent
1,000,000 or more	10	76.9%	0	0.0%	3	23.1%	0	0.0%	13	100.0%
500,000 to 999,999	24	63.2	8	21.1	4	10.5	2	5.3	38	100.0
250,000 to 499,999	24	42.1	31	54.4	2	3.6	0	0.0	57	100.0
100,000 to 249,999	53	28.9	94	51.0	32	17.4	5	2.7	184	100.0
50,000 to 99,999	81	20.0	232	57.3	78	19.3	14	3.4	405	100.0
25,000 to 49,999	125	14.5	467	54.4	220	25.6	47	5.4	860	100.0
10,000 to 24,999	291	13.9	1,151	55.0	576	27.6	73	3.5	2,090	100.0
5,000 to 9,999	431	17.4	1,353	54.7	632	25.5	58	2.4	2,475	100.0
2,500 to 4,999	509	18.3	1,528	54.8	693	24.8	59	2.1	2,787	100.0
Under 2,500	1,264	21.2	3,312	55.6	1,312	22.0	72	1.2	5,961	100.0
Total	2,806	18.9	8,177	55.0	3,553	23.9	329	2.2	14,872	100.0

Source: FEMA US Fire Administration 2002
Survey of the Needs of the US Fire Service

The above projections are based on 5,083 departments reporting yes to Question 36a and also reporting on Question 36c. Numbers may not add to totals due to rounding.

Q. 36c: If [technical rescue and EMS for a building with 50 occupants after structural collapse is within your department's scope], how far would you have to go to obtain enough specialized equipment to handle this incident?

Table 54
For Departments Where Technical Rescue and EMS for a Building With 50 Occupants After Structural Collapse Is Within Their Scope, Do They Have a Plan for Working With Others?
by Community Size
(Q. 36d)

Population of Community	Yes – Written Agreement		Yes – Informal		Yes – Other		No		Total	
	Number Depts	Percent	Number Depts	Percent	Number Depts	Percent	Number Depts	Percent	Number Depts	Percent
1,000,000 or more	12	92.3%	1	7.6%	0	0.0%	0	0.0%	13	100.0%
500,000 to 999,999	22	57.8	13	34.2	3	7.9	0	0.0	38	100.0
250,000 to 499,999	43	75.4	14	24.5	0	0.0	0	0.0	57	100.0
100,000 to 249,999	108	58.7	53	28.8	12	6.5	12	6.5	184	100.0
50,000 to 99,999	227	56.0	125	30.9	14	3.4	38	9.4	405	100.0
25,000 to 49,999	376	43.7	293	34.0	57	6.6	134	15.7	860	100.0
10,000 to 24,999	851	40.7	793	37.9	113	5.4	333	16.0	2,090	100.0
5,000 to 9,999	824	33.3	1,057	42.7	148	6.0	444	17.9	2,475	100.0
2,500 to 4,999	799	28.7	1,218	43.7	147	5.3	622	22.3	2,787	100.0
Under 2,500	1,686	28.3	2,429	40.8	406	6.8	1,441	24.2	5,961	100.0
Total	4,941	33.2	6,000	40.4	900	6.1	3,025	20.3	14,872	100.0

Source: FEMA US Fire Administration 2002
Survey of the Needs of the US Fire Service

The above projections are based on 5,090 departments reporting yes to Question 36a and also reporting on Question 36d. Numbers may not add to totals due to rounding.

Q. 36d: Do you have a plan for working on others on [technical rescue and EMS for a building with 50 occupants after structural collapse]?

113

Table 55
Is a Hazmat and EMS Incident Involving Chemical/Biological Agents
and 10 Injuries Within the Scope of Department?
by Community Size
(Q. 37a)

Population	Yes		No		Total	
	Number		Number		Number	
of Community	Depts	Percent	Depts	Percent	Depts	Percent
1,000,000 or more	13	100.0%	0	0.0%	13	100.0%
500,000 to 999,999	38	100.0	0	0.0	38	100.0
250,000 to 499,999	63	98.4	1	1.6	64	100.0
100,000 to 249,999	200	93.0	15	7.0	215	100.0
50,000 to 99,999	447	91.8	40	8.2	487	100.0
25,000 to 49,999	889	84.4	164	15.6	1,053	100.0
10,000 to 24,999	2,183	76.8	670	23.6	2,843	100.0
5,000 to 9,999	2,486	68.5	1,143	31.5	3,629	100.0
2,500 to 4,999	2,739	59.9	1,833	40.1	4,572	100.0
Under 2,500	6,295	46.8	7,145	53.2	13,440	100.0
Total	15,352	58.3	11,002	41.7	26,354	100.0

Source: FEMA US Fire Administration 2002
        Survey of the Needs of the US Fire Service

The above projections are based on 8,250 departments reporting on Question 37a. Numbers may not add to totals due to rounding.

Q. 37a: Is [hazmat and EMS for an incident involving chemical/biological agents and 10 injuries] within your department's scope?

Table 56
For Departments Where a Hazmat and EMS Incident
Involving Chemical/Biological Agents and 10 Injuries Is Within Their Scope
How Far Do They Have to Go to Obtain Sufficient People
With Specialized Training to Handle Such an Incident?
by Community Size
(Q. 37b)

Population of Community	Local Number Depts	Local Percent	Regional Number Depts	Regional Percent	State Number Depts	State Percent	National Number Depts	National Percent	Total Number Depts	Total Percent
1,000,000 or more	13	100.0%	0	0.0%	0	0.0%	0	0.0%	13	100.0%
500,000 to 999,999	34	89.5	2	5.3	2	5.3	0	0.0	38	100.0
250,000 to 499,999	47	74.6	11	17.5	5	7.9	0	0.0	63	100.0
100,000 to 249,999	121	60.5	72	36.0	6	3.0	1	0.5	200	100.0
50,000 to 99,999	159	35.6	234	52.3	47	10.5	7	1.6	447	100.0
25,000 to 49,999	283	31.8	463	52.1	130	14.6	12	1.3	889	100.0
10,000 to 24,999	503	23.0	1,180	54.1	467	21.4	32	1.5	2,182	100.0
5,000 to 9,999	522	21.0	1,347	54.2	569	22.8	47	1.9	2,486	100.0
2,500 to 4,999	497	18.1	1,456	53.2	733	26.8	53	1.9	2,739	100.0
Under 2,500	1,251	19.9	3,068	48.7	1,884	29.9	92	1.5	6,295	100.0
Total	3,427	22.3	7,832	51.0	3,843	25.0	244	1.6	15,352	100.0

Source: FEMA US Fire Administration 2002
Survey of the Needs of the US Fire Service

The above projections are based on 5,292 departments reporting yes to Question 37a and also reporting on Question 37b.
Numbers may not add to totals due to rounding.

Q. 37b: If [hazmat and EMS for an incident involving chemical/biological agents and 10 injuries is within your department's scope],
how far would you have to go to obtain enough people with specialized training for this incident?

115

Table 57
For Departments Where a Hazmat and EMS Incident Involving Chemical/Biological Agents and 10 Injuries Is Within Their Scope How Far Do They Have to Go to Obtain Sufficient Specialized Equipment to Handle Such An Incident? by Community Size
(Q. 37c)

Population of Community	Local		Regional		State		National		Total	
	Number Depts	Percent	Number Depts	Percent	Number Depts	Percent	Number Depts	Percent	Number Depts	Percent
1,000,000 or more	12	92.3%	1	7.7%	0	0.0%	0	0.0%	13	100.0%
500,000 to 999,999	31	81.6	4	10.5	2	5.2	1	2.6	38	100.0
250,000 to 499,999	44	69.8	11	17.5	7	11.1	1	1.6	63	100.0
100,000 to 249,999	109	54.5	76	38.0	14	7.0	1	0.5	200	100.0
50,000 to 99,999	142	31.8	224	50.1	73	16.3	8	1.8	447	100.0
25,000 to 49,999	226	25.4	467	52.5	177	19.9	17	1.9	889	100.0
10,000 to 24,999	446	20.4	1,155	53.0	536	24.6	45	2.1	2,182	100.0
5,000 to 9,999	424	17.1	1,337	53.8	661	26.6	64	2.6	2,486	100.0
2,500 to 4,999	416	15.2	1,421	51.9	823	30.1	77	2.8	2,739	100.0
Under 2,500	1,014	16.1	2,992	47.5	2,184	34.7	103	1.7	6,295	100.0
Total	2,862	18.7	7,683	50.1	4,478	29.2	317	2.1	15,352	100.0

Source: FEMA US Fire Administration 2002
Survey of the Needs of the US Fire Service

The above projections are based on 5,239 departments reporting yes to Question 37a and also reporting on Question 37c. Numbers may not add to totals due to rounding.

Q. 37c: If [hazmat and EMS for an incident involving chemical/biological agents and 10 injuries is within your department's scope], how far would you have to go to obtain enough specialized equipment to handle this incident?

116

Table 58
For Departments Where a Hazmat and EMS Incident Involving Chemical/Biological Agents and 10 Injuries Is Within Their Scope Do They Have a Plan for Working With Others?
by Community Size
(Q. 37d)

Population of Community	Yes – Written Agreement		Yes – Informal		Yes – Other		No		Total	
	Number Depts	Percent	Number Depts	Percent	Number Depts	Percent	Number Depts	Percent	Number Depts	Percent
1,000,000 or more	12	92.3%	1	7.7%	0	0.0%	0	0.0%	13	100.0%
500,000 to 999,999	27	71.1	8	21.1	3	7.9	0	0.0	38	100.0
250,000 to 499,999	43	68.3	17	27.0	3	4.8	0	0.0	63	100.0
100,000 to 249,999	138	69.0	45	22.5	12	6.0	5	2.5	200	100.0
50,000 to 99,999	304	68.0	101	22.6	19	4.3	23	5.1	447	100.0
25,000 to 49,999	538	60.6	241	27.1	47	5.3	61	6.9	889	100.0
10,000 to 24,999	1,031	47.2	858	39.3	118	5.4	175	8.0	2 183	100.0
5,000 to 9,999	911	36.7	1,103	44.3	148	6.0	323	13.0	2,486	100.0
2,500 to 4,999	811	29.6	1,245	45.5	192	7.1	486	17.8	2,739	100.0
Under 2,500	1,738	27.6	2,780	44.2	387	6.2	1,388	22.0	6,295	100.0
Total	5,552	36.2	6,400	41.7	929	6.0	2,461	16.0	15 352	100.0

Source: FEMA US Fire Administration 2002
Survey of the Needs of the US Fire Service

The above projections are based on 5,227 departments reporting yes to Question 37a and also reporting on Question 37d. Numbers may not add to totals due to rounding.

Q. 37d: Do you have a plan for working on others on [hazmat and EMS for an incident involving chemical/biological agents and 10 injuries]?

117

Table 59
Is a Wildland/Urban Interface Fire Affecting 500 Acres
Within the Scope of Department?
by Community Size
(Q. 38a)

Population	Yes		No		Total	
	Number		Number		Number	
of Community	Depts	Percent	Depts	Percent	Depts	Percent
1,000,000 or more	9	69.2%	4	30.8%	13	100.0%
500,000 to 999,999	30	78.9	8	21.1	38	100.0
250,000 to 499,999	39	60.9	25	39.1	64	100.0
100,000 to 249,999	131	60.9	84	39.1	215	100.0
50,000 to 99,999	278	57.1	209	42.9	487	100.0
25,000 to 49,999	551	52.3	502	47.7	1,053	100.0
10,000 to 24,999	1,735	61.0	1,109	39.0	2,843	100.0
5,000 to 9,999	2,491	68.6	1,138	31.4	3,629	100.0
2,500 to 4,999	3,279	71.7	1,293	28.3	4,572	100.0
Under 2,500	9,699	72.2	3,741	27.8	13,440	100.0
Total	18,242	69.2	8,112	30.8	26,354	100.0

Source: FEMA US Fire Administration 2002
Survey of the Needs of the US Fire Service

The above projections are based on 8,248 departments reporting on Question 38a. Numbers may not add to totals due to rounding.

Q. 38a: Is [a wildland/urban interface fire affecting 500 acres] within your department's scope?

Table 60
For Departments Where a Wildland/Urban
Interface Fire Affecting 500 Acres Is Within Their Scope
How Far Do They Have to Go to Obtain Sufficient People
With Specialized Training to Handle Such an Incident?
by Community Size
(Q. 38b)

Population of Community	Local Number Depts	Local Percent	Regional Number Depts	Regional Percent	State Number Depts	State Percent	National Number Depts	National Percent	Total Number Depts	Total Percent
1,000,000 or more	4	44.4%	3	33.3%	2	22.2%	0	0.0%	9	100.0%
500,000 to 999,999	14	46.7	10	33.3	6	20.0	0	0.0	30	100.0
250,000 to 499,999	16	41.0	20	51.3	3	7.7	0	0.0	39	100.0
100,000 to 249,999	39	29.8	62	47.3	28	21.4	2	1.5	131	100.0
50,000 to 99,999	80	28.8	139	50.0	60	21.6	1	0.4	278	100.0
25,000 to 49,999	149	27.0	266	48.3	131	23.8	5	0.9	551	100.0
10,000 to 24,999	492	28.4	743	42.8	486	28.0	14	0.8	1,735	100.0
5,000 to 9,999	775	31.1	1,034	41.5	649	26.0	34	1.4	2,491	100.0
2,500 to 4,999	1,104	33.7	1,320	40.3	803	24.5	47	1.4	3,279	100.0
Under 2,500	4,053	41.8	3,801	39.2	1,700	17.5	143	1.5	9,699	100.0
Total	6,722	36.9	7,399	40.6	3,868	21.2	247	1.4	18,242	100.0

Source: FEMA US Fire Administration 2002
Survey of the Needs of the US Fire Service

The above projections are based on 5,503 departments reporting yes to Question 38a and also reporting on Question 38b. Numbers may not add to totals due to rounding.

Q. 38b: If [wildland/urban interface fire affecting 500 acres is within your department's scope], how far would you have to go to obtain enough people with specialized training for this incident?

119

Table 61
For Departments Where a Wildland/Urban
Interface Fire Affecting 500 Acres Is Within Their Scope
How Far Do They Have to Go to Obtain Sufficient
Specialized Equipment to Handle Such An Incident?
by Community Size
(Q. 38c)

Population of Community	Local		Regional		State		National		Total	
	Number Depts	Percent	Number Depts	Percent	Number Depts	Percent	Number Depts	Percent	Number Depts	Percent
1,000,000 or more	4	44.4%	3	33.3%	2	22.2%	0	0.0%	9	100.0%
500,000 to 999,999	10	33.3	14	46.7	6	20.0	0	0.0	30	100.0
250,000 to 499,999	15	38.5	19	48.7	5	12.8	0	0.0	39	100.0
100,000 to 249,999	32	24.4	63	48.1	34	25.9	2	1.5	131	100.0
50,000 to 99,999	71	25.5	143	51.4	62	22.3	1	0.3	278	100.0
25,000 to 49,999	131	23.8	257	46.6	159	28.9	4	0.7	551	100.0
10,000 to 24,999	449	25.9	723	41.7	542	31.2	20	1.2	1,735	100.0
5,000 to 9,999	644	25.9	1,088	43.7	719	28.8	40	1.6	2,491	100.0
2,500 to 4,999	986	30.1	1,330	40.6	913	27.8	50	1.5	3,279	100.0
Under 2,500	3,534	36.4	3,902	40.2	2,092	21.6	170	1.8	9,699	100.0
Total	5,876	32.2	7,538	41.3	4,534	24.9	287	1.6	18,242	100.0

Source: FEMA US Fire Administration 2002
Survey of the Needs of the US Fire Service

The above projections are based on 5,468 departments reporting yes to Question 38a and also reporting on Question 38c. Numbers may not add to totals due to rounding.

Q. 38c: If [wildland/urban interface fire affecting 500 acres is within your department's scope], how far would you have to go to obtain enough specialized equipment to handle this incident?

Table 62
For Departments Where a Wildland/Urban
Interface Fire Affecting 500 Acres Is Within Their Scope
Do They Have a Plan for Working With Others?
by Community Size
(Q. 38d)

Population of Community	Yes – Written Agreement		Yes – Informal		Yes – Other		No		Total	
	Number Depts	Percent	Number Depts	Percent	Number Depts	Percent	Number Depts	Percent	Number Depts	Percent
1,000,000 or more	9	100.0%	0	0.0%	0	0.0%	0	0.0%	9	100.0%
500,000 to 999,999	19	63.3	7	23.3	1	3.3	3	10.0	30	100.0
250,000 to 499,999	27	69.2	7	17.9	3	7.7	2	5.1	39	100.0
100,000 to 249,999	93	71.0	30	22.9	3	2.3	5	3.8	131	100.0
50,000 to 99,999	195	70.1	52	18.7	12	4.3	19	6.8	278	100.0
25,000 to 49,999	357	64.8	142	25.8	17	3.1	35	6.3	551	100.0
10,000 to 24,999	945	54.5	583	33.5	73	4.2	132	7.5	1,735	100.0
5,000 to 9,999	1,195	48.0	977	39.2	120	4.8	197	7.9	2,491	100.0
2,500 to 4,999	1,451	44.3	1,390	42.4	153	4.7	285	8.7	3,279	100.0
Under 2,500	4,301	44.3	4,136	42.6	522	5.4	740	7.6	9,699	100.0
Total	8,590	47.1	7,324	40.2	905	5.0	1,418	7.8	18,242	100.0

Source: FEMA US Fire Administration 2002
Survey of the Needs of the US Fire Service

The above projections are based on 5,448 departments reporting yes to Question 38a and also reporting on Question 38d.
Numbers may not add to totals due to rounding.

Q. 38d: Do you have a plan for working on others on [wildland/urban interface fire affecting 500 acres]?

Table 63
Is Mitigation of a Developing Major Flood
Within the Scope of Department?
by Community Size
(Q. 39a)

Population	Yes		No		Total	
	Number		Number		Number	
of Community	Depts	Percent	Depts	Percent	Depts	Percent
1,000,000 or more	9	69.2%	4	30.8%	13	100.0%
500,000 to 999,999	26	68.4	12	31.6	38	100.0
250,000 to 499,999	42	65.6	22	34.4	64	100.0
100,000 to 249,999	138	64.2	77	35.8	215	100.0
50,000 to 99,999	300	61.6	187	38.4	487	100.0
25,000 to 49,999	616	58.5	437	41.5	1,053	100.0
10,000 to 24,999	1,540	54.2	1,303	45.8	2,843	100.0
5,000 to 9,999	1,908	52.6	1,721	47.4	3,629	100.0
2,500 to 4,999	2,239	49.0	2,333	51.0	4,572	100.0
Under 2,500	5,194	38.6	8,246	61.4	13,440	100.0
Total	12,014	45.6	14,340	54.4	26,354	100.0

Source: FEMA US Fire Administration 2002
       Survey of the Needs of the US Fire Service

The above projections are based on 8,162 departments reporting yes on Question 39a. Numbers may not add to totals due to rounding.

Q. 39a: Is [mitigation (confining, slowing, etc.) of a developing major flood] within your department's scope?

Table 64
For Departments Where Mitigation of a Major Flood Is Within Their Scope
How Far Do They Have to Go to Obtain Sufficient People
With Specialized Training to Handle Such an Incident?
by Community Size
(Q. 39b)

Population of Community	Local		Regional		State		National		Total	
	Number Depts	Percent	Number Depts	Percent	Number Depts	Percent	Number Depts	Percent	Number Depts	Percent
1,000,000 or more	7	77.7%	1	11.1%	1	11.1%	0	0.0%	9	100.0%
500,000 to 999,999	15	57.7	8	30.8	3	11.5	0	0.0	26	100.0
250,000 to 499,999	12	28.6	15	35.7	15	35.7	0	0.0	42	100.0
100,000 to 249,999	31	22.5	57	41.3	45	32.6	5	3.6	138	100.0
50,000 to 99,999	53	17.7	133	44.3	109	36.3	5	1.7	300	100.0
25,000 to 49,999	128	20.8	238	38.6	221	35.9	28	4.5	616	100.0
10,000 to 24,999	318	20.6	577	37.5	600	38.9	45	2.9	1,540	100.0
5,000 to 9,999	432	22.6	705	37.0	719	37.7	51	2.7	1,908	100.0
2,500 to 4,999	648	29.0	850	38.0	703	31.3	39	1.7	2,239	100.0
Under 2,500	1,578	30.4	2,013	38.7	1,510	29.1	93	1.8	5,194	100.0
Total	3,222	26.8	4,598	38.3	3,925	32.7	266	2.2	14,014	100.0

Source: FEMA US Fire Administration 2002
Survey of the Needs of the US Fire Service

The above projections are based on 3,967 departments reporting yes to Question 39a and also reporting on Question 39b.
Numbers may not add to totals due to rounding.

Q. 39b: If [mitigation (confining, slowing, etc.) of a developing major flood is within your department's scope], how far would you have to go to obtain enough people with specialized training for this incident?

Table 65
For Departments Where Mitigation of a Major Flood Is Within Their Scope
How Far Do They Have to Go to Obtain Sufficient
Specialized Equipment to Handle Such An Incident?
by Community Size
(Q. 39c)

Population of Community	Local		Regional		State		National		Total	
	Number Depts	Percent	Number Depts	Percent	Number Depts	Percent	Number Depts	Percent	Number Depts	Percent
1,000,000 or more	3	33.3%	5	55.5%	1	11.1%	0	0.0%	9	100.0%
500,000 to 999,999	12	46.2	9	34.6	5	19.2	0	0.0	26	100.0
250,000 to 499,999	11	26.2	16	38.1	14	33.3	1	2.4	42	100.0
100,000 to 249,999	24	17.4	65	47.1	43	31.2	6	4.3	138	100.0
50,000 to 99,999	42	14.0	128	42.7	124	41.3	7	2.3	300	100.0
25,000 to 49,999	105	17.0	234	38.0	245	39.8	32	5.2	616	100.0
10,000 to 24,999	280	18.2	587	38.2	610	39.6	62	4.0	1,540	100.0
5,000 to 9,999	367	19.2	715	37.9	755	40.0	56	3.0	1,908	100.0
2,500 to 4,999	570	25.4	810	35.8	816	36.1	59	2.7	2,239	100.0
Under 2,500	1,363	26.2	2,026	39.0	1,673	32.2	130	2.5	5,194	100.0
Total	2,776	23.1	4,596	38.3	4,276	35.6	353	2.9	12,014	100.0

Source: FEMA US Fire Administration 2002
Survey of the Needs of the US Fire Service

The above projections are based on 3,947 departments reporting yes to Question 39a and also reporting on Question 39c.
Numbers may not add to totals due to rounding.

Q. 39c: If [mitigation (confining, slowing, etc.) of a developing major flood is within your department's scope], how far would you have to go to obtain enough specialized equipment to handle this incident?

Table 66
For Departments Where Mitigation of a Major Flood Is Within Their Scope
Do They Have a Plan for Working With Others?
by Community Size
(Q. 39d)

Population of Community	Yes – Written Agreement		Yes – Informal		Yes – Other		No		Total	
	Number Depts	Percent	Number Depts	Percent	Number Depts	Percent	Number Depts	Percent	Number Depts	Percent
1,000,000 or more	4	44.4%	3	33.3%	1	11.1%	1	11.1%	9	100.0%
500,000 to 999,999	12	46.2	12	46.2	2	7.7	0	0.0	26	100.0
250,000 to 499,999	25	59.5	13	31.0	3	7.1	1	2.4	42	100.0
100,000 to 249,999	77	55.8	38	27.5	10	7.2	13	9.4	138	100.0
50,000 to 99,999	174	58.0	76	25.3	11	3.7	38	12.7	300	100.0
25,000 to 49,999	265	43.1	184	30.0	35	5.7	132	21.5	616	100.0
10,000 to 24,999	554	36.0	572	37.1	70	4.5	344	22.4	1,540	100.0
5,000 to 9,999	588	30.8	778	40.7	85	4.5	456	23.9	1,908	100.0
2,500 to 4,999	582	26.0	957	42.8	120	5.4	579	25.8	2,239	100.0
Under 2,500	1,084	20.9	2,380	45.8	266	5.1	1,460	28.2	5,194	100.0
Total	3,365	28.0	5,011	41.7	603	5.0	3,025	25.2	12,014	100.0

Source: FEMA US Fire Administration 2002
Survey of the Needs of the US Fire Service

The above projections are based on 3,916 departments reporting yes to Question 39a and also reporting on Question 39d. Numbers may not add to totals due to rounding.

Q. 39d: Do you have a plan for working on others on [mitigation (confining, slowing, etc.) of a developing major flood]?

NEW AND EMERGING TECHNOLOGY

Tables 67-70 (pp. 128-131) address the ownership and planned purchase of four types of relatively new technologies.

One quarter of fire departments now own thermal imaging cameras, but most that do not already have them – and two-fifths of all departments – have no plans to acquire them. (See Table 67.)

Only one department in 28 has mobile data terminals, though most of the fire departments protecting at least 500,000 population have them, and most departments (78% overall) have no plans to acquire them. (See Table 68.)

Only one department in 50 has advanced personnel location equipment, though one-fifth of the fire departments protecting communities of at least 500,000 population have them. Plans to acquire them vary considerably by department size, but four-fifths of departments overall have no plans. The survey did not provide details on what constituted advanced personnel location equipment, which raises the possibility that departments differed in their views of the kind of equipment that would qualify as such. (See Table 69.)

Only one department in 23 has equipment to collect chemical or biological samples for remote analysis, though most of the fire departments protecting communities of at least 250,000 population have such equipment. Only one department in 12 overall has plans to acquire such equipment. (See Table 70.)

Table 67
Plans to Acquire Thermal Imaging Cameras
by Community Size
(Q. 40)

Population of Community	Now Own One		Plan to Have One in One Year		Plan to Have One in Five Years		No Plans to Acquire		Total	
	Number Depts	Percent	Number Depts	Percent	Number Depts	Percent	Number Depts	Percent	Number Depts	Percent
1,000,000 or more	10	76.9%	3	23.1%	0	0.0%	0	0.0%	13	100.0%
500,000 to 999,999	35	92.1	1	2.6	2	5.3	0	0.0	38	100.0
250,000 to 499,999	51	81.4	9	14.0	2	2.3	2	2.3	64	100.0
100,000 to 249,999	174	80.9	26	12.1	9	4.1	6	2.9	215	100.0
50,000 to 99,999	370	76.0	59	12.0	33	6.8	25	5.2	487	100.0
25,000 to 49,999	719	68.3	132	12.6	134	12.7	68	6.4	1,053	100.0
10,000 to 24,999	1,622	57.0	356	12.5	533	18.7	333	11.7	2,843	100.0
5,000 to 9,999	1,402	38.6	452	12.4	986	27.2	790	21.8	3,629	100.0
2,500 to 4,999	929	20.3	443	9.7	1,487	32.5	1,714	37.5	4,572	100.0
Under 2,500	1,124	8.4	578	4.3	3,105	23.1	8,632	64.2	13,440	100.0
Total	6,435	24.4	2,058	7.8	6,290	23.9	11,570	43.9	26,354	100.0

Source: FEMA US Fire Administration 2002
Survey of the Needs of the US Fire Service

The above projections are based on 8,355 departments reporting on Question 40. Numbers may not add to totals due to rounding.

Q. 40: Do you have any [thermal imaging cameras] now or plan to acquire any?

128

Table 68
Plans to Acquire Mobile Data Terminals
by Community Size
(Q. 41)

Population of Community	Now Own One		Plan to Have One in One Year		Plan to Have One in Five Years		No Plans to Acquire		Total	
	Number Depts	Percent	Number Depts	Percent	Number Depts	Percent	Number Depts	Percent	Number Depts	Percent
1,000,000 or more	11	84.6%	2	15.4%	0	0.0%	0	0.0%	13	100.0%
500,000 to 999,999	23	60.6	1	2.6	10	26.3	4	10.5	38	100.0
250,000 to 499,999	25	39.1	11	17.2	17	26.6	11	17.2	64	100.0
100,000 to 249,999	63	29.3	51	23.7	62	28.9	39	17.9	215	100.0
50,000 to 99,999	115	23.4	65	13.4	168	34.5	139	28.5	487	100.0
25,000 to 49,999	152	14.4	126	12.0	406	38.6	369	35.0	1,053	100.0
10,000 to 24,999	217	7.6	249	8.8	917	32.3	1,459	51.3	2,843	100.0
5,000 to 9,999	137	3.8	150	4.1	815	22.4	2,527	69.6	3,629	100.0
2,500 to 4,999	87	1.9	96	2.1	635	13.9	3,754	82.1	4,572	100.0
Under 2,500	133	1.0	148	1.1	913	6.8	12,246	91.1	13,440	100.0
Total	960	3.6	899	3.4	3,944	15.0	20,551	78.0	26,354	100.0

Source: FEMA US Fire Administration 2002
Survey of the Needs of the US Fire Service

The above projections are based on 8,300 departments reporting on Question 41. Numbers may not add to totals due to rounding.

Q. 41: Do you have any [mobile data terminals] now or plan to acquire any?

Table 69
Plans to Acquire Advanced Personnel Location Equipment
by Community Size
(Q. 42)

Population of Community	Now Own One		Plan to Have One in One Year		Plan to Have One in Five Years		No Plans to Acquire		Total	
	Number Depts	Percent	Number Depts	Percent	Number Depts	Percent	Number Depts	Percent	Number Depts	Percent
1,000,000 or more	3	23.1%	4	30.8%	5	38.5%	1	7.7%	13	100.0%
500,000 to 999,999	8	21.1	1	2.6	13	34.2	4	10.5	38	100.0
250,000 to 499,999	3	4.7	3	4.7	27	42.2	31	48.4	64	100.0
100,000 to 249,999	19	8.9	19	8.9	61	28.4	116	53.5	215	100.0
50,000 to 99,999	19	3.9	23	4.8	147	30.1	298	61.1	487	100.0
25,000 to 49,999	37	3.5	54	5.1	305	29.0	657	62.4	1,053	100.0
10,000 to 24,999	86	3.0	94	3.3	626	22.0	2,037	71.7	2,843	100.0
5,000 to 9,999	84	2.3	71	2.0	679	18.7	2,795	77.0	3,629	100.0
2,500 to 4,999	91	2.0	94	2.1	594	13.0	3,793	83.0	4,572	100.0
Under 2,500	170	1.3	144	1.1	1,493	11.1	11,633	86.6	13,440	100.0
Total	523	2.0	510	1.9	3,951	15.0	21,371	81.1	26,354	100.0

Source: FEMA US Fire Administration 2002
Survey of the Needs of the US Fire Service

The above projections are based on 8,181 departments reporting on Question 42. Numbers may not add to totals due to rounding.

Q. 42: Do you have any [advanced personnel location equipment] now or plan to acquire any?

Table 70
Plans to Acquire Equipment to Collect Chemical/Biological Samples for Analysis Elsewhere by Community Size
(Q. 43)

Population of Community	Now Own One		Plan to Have One in One Year		Plan to Have One in Five Years		No Plans to Acquire		Total	
	Number Depts	Percent	Number Depts	Percent	Number Depts	Percent	Number Depts	Percent	Number Depts	Percent
1,000,000 or more	12	90.0%	0	0.0%	1	10.0%	0	0.0%	13	100.0%
500,000 to 999,999	32	84.9	0	0.0	4	9.1	2	6.1	38	100.0
250,000 to 499,999	43	67.4	7	11.6	6	9.5	7	11.6	64	100.0
100,000 to 249,999	103	47.8	39	18.2	22	10.3	51	23.6	215	100.0
50,000 to 99,999	139	28.5	93	19.0	34	7.0	222	45.5	487	100.0
25,000 to 49,999	209	19.9	87	8.3	134	12.7	623	59.1	1,053	100.0
10,000 to 24,999	256	9.0	187	6.6	306	10.8	2,093	73.6	2,843	100.0
5,000 to 9,999	115	3.2	120	3.3	293	8.1	3,101	85.5	3,629	100.0
2,500 to 4,999	93	2.0	70	1.5	256	5.6	4,154	90.9	4,572	100.0
Under 2,500	127	1.0	123	0.9	481	3.6	12,709	94.6	13,440	100.0
Total	1,125	4.3	726	2.8	1,533	5.8	22,970	87.2	26,354	100.0

Source: FEMA US Fire Administration 2002
Survey of the Needs of the US Fire Service

The above projections are based on 8,250 departments reporting on Question 43. Numbers may not add to totals due to rounding.

Q. 43: Do you have any [equipment to collect chem/bio samples for analysis elsewhere] now or plan to acquire any?

131

APPENDIX 1: SURVEY METHODOLOGY

The Fire Service Needs Assessment Survey was conducted as a census, with appropriate adjustments for non-response. The choice of a census approach rather than a random sample approach was based on two considerations.

First, the survey is a specific requirement of PL 106-398 in Section 1701, Sec. 33(b), and the larger act is designed to provide the U.S. Fire Service with appropriate assistance for their legitimate needs. Given this intended application, there was general agreement that fire departments would view the survey as an opportunity rather than a burden, an opportunity that every department would wish to be given.

Second, current usage of some of the types of equipment and training to be addressed in the survey was believed to be sufficiently rare that the study would need the largest possible base for analysis.

The NFPA used its own list of local fire departments as the mailing list and sampling frame of all fire departments in the U.S. In all, 26,354 fire departments were mailed survey forms.

The content of the survey was developed by NFPA, in collaboration with an ad hoc technical advisory group consisting of representatives of the full spectrum of national organizations and related disciplines associated with the management of fire and related hazards and risks in the U.S. A copy of the survey form is provided at the end of this section of the report.

The fire departments were mailed the survey form the week of December 11, 2001. Despite being mailed during the holiday season, the first mailing resulted in a very strong initial response from departments. A second mailing was sent the week of February 11 to departments that had not responded to the initial mailing.

Overall, NFPA received 12,240 completed surveys and has edited, coded, and keyed 8,416 surveys for analysis in this report. The overall response rate is 46%, which is unusually high for a survey involving a large number of smaller departments. The better-than-expected response is due in part to the subject of the survey, its intended use, and undoubtedly the events of September 11.

With such a large number of surveys returned and the deadline to have the final report ready by September 30, priority was given to editing and keying forms from larger departments – those protecting populations of 50,000 or more. An attempt was made to edit and key all the returned forms from these larger departments. However, we also keyed a sufficient sample size to make accurate estimates of smaller communities as well. The NFPA included all survey forms it had edited and keyed by August 22 as the basis for this report.

At USFA request, NFPA prepared two preliminary reports in May and June. These reports were based on the 5,100 surveys that had been edited and keyed at that point. Those results have been compared to the results for this report and found to be very similar. It is therefore believed that the surveys being keyed late would not, if analyzed, materially affect the results, either nationally or by community size. However, they will permit a much larger share of US fire departments to have participated, in what clearly is shaping up as the highest-participation and most-detailed database on fire service resources and needs ever assembled.

Because a census was undertaken, there is no sampling error. However, response rates varied considerably by community size, with departments protecting smaller communities responding at a much lower rate than departments protecting larger communities (Table A-1). Thus where national total results are presented, adjustments were made to account for variation in response rates by community size.

Table A-1
Fire Departments Responding to Fire Department Needs Survey
by Size of Community

Population of Community	Number of Surveys Mailed	Percent of Surveys Responded	Of Surveys Received, Percent Keyed As of Cutoff Date	Number of Surveys Received and Keyed as of Cutoff Date
1,000,000 or more	13	77%	100%	10
500,000 to 999,999	38	87%	100%	33
250,000 to 499,999	64	75%	100%	48
100,000 to 249,999	215	83%	100%	181
50,000 to 99,999	487	78%	100%	380
25,000 to 49,999	1,053	76%	83%	624
10,000 to 24,999	2,843	65%	77%	1,442
5,000 to 9,999	3,629	56%	70%	1,412
2,500 to 4,999	4,572	52%	67%	1,599
Under 2,500	13,440	34%	59%	2,687
Total	26,354	46%	69%	8,416

Source: FEMA U.S. Fire Administration 2002
          Survey of the Needs of the U.S. Fire Service

Numbers may not add to totals due to rounding.

APPENDIX 2: SURVEY FORM

The next four pages contain the Needs Assessment Survey form.

It was printed on legal size paper (8-1/2" x 14") but has been shrunk to fit letter size paper here.

FEDERAL EMERGENCY MANAGEMENT AGENCY
U.S. FIRE ADMINISTRATION
SURVEY OF THE NEEDS OF THE U.S. FIRE SERVICE

PART I. IDENTIFYING INFORMATION

Name of person completing form: Date:

Title of person completing form:

Non emergency phone number: (   )   Fax: (   )

e mail address:

Please use enclosed postpaid envelope and return completed form to:

Fire Analysis and Research Division
1 Batterymarch Park
Quincy, MA 02269 9101 USA
Fax: (617) 984 7478

If you fax the form back, please reduce it first to 8 1/2" x 11" size.

PART II. BASIC INFORMATION

1. Population (Number of permanent residents) your department has *primary* responsibility
to protect (exclude mutual aid areas):

2. Area (in square miles) your department has primary responsibility
to protect (exclude mutual aid areas):

PART III. BUDGET INFORMATION

3. Do you have a plan for apparatus replacement on a regular schedule?

❏ Yes ❏ No

4. Does your normal budget cover the costs of apparatus replacement?

❏ Yes, budget covers costs

❏ No, must raise funds or seek special appropriation for purchase

(Questions 5 and 6 are for all or mostly volunteer or call departments ONLY. Indicate % for each, so percents sum to 100 for each question):

5. What share (%) of your budgeted revenue is from: Fire district or other taxes

   Payments per call   Other local payments   State government

   Fund raising (e.g., donations, raffles, suppers, events)

   Other (specify)

6. What share (%) of your apparatus was: Purchased new   Donated new

   Purchased used   Donated used

   Converted vehicles not designed as FD apparatus

   Other (specify)

PART IV. PERSONNEL AND THEIR CAPABILITIES

7. Total number of full-time (career) uniformed fire fighters:

8. Total number of active part-time (call or volunteer) fire fighters:

9. Average number of career/paid firefighters on duty available to respond to emergencies
(total number for department):

10. Average number of call/volunteer personnel who respond to a mid-day house fire:

11. Number of on-duty career/paid personnel assigned to an engine/pumper (Circle one)

            1 2   3   4   5+   Not applicable

12. Number of on duty career/paid personnel assigned to a ladder/aerial (Circle one)

            1 2   3   4   5+   Not applicable

PART IV. PERSONNEL AND THEIR CAPABILITIES (continued)

13. Structural firefighting.

a. Is this a role your department performs? (Check one)　　❏ Yes　　❏ No

b. If yes, how many of your personnel who perform this duty have received formal training (not just on the job)?
(Check one)　❏ All　❏ Most　❏ Some　❏ None

c. Have any of your personnel been certified to any of the following levels?
(Circle letters for all that apply)　A. Firefighter Level I　　B. Firefighter Level II

14. Emergency medical service (EMS).

a. Is this a role your department performs? (Check one)　　❏ Yes　　❏ No

b. If yes, how many of your personnel who perform this duty have received formal training (not just on the job)?
(Check one)　❏ All　❏ Most　❏ Some　❏ None

c. If yes to a, have any of your personnel been certified to any of the following levels?
(Circle letters for all that apply)

A. First responder　　B. Basic Life Support (BLS)/EMT Intermediate (EMT I)

C. Advanced Life Support (ALS)/EMT Intermediate (EMT I)　D. ALS/Paramedic

15. Hazardous materials response (Hazmat).

a. Is this a role your department performs? (Check one)　　❏ Yes　　❏ No

b. If yes, how many of your personnel who perform this duty have received formal training (not just on the job)?
(Check one)　❏ All　❏ Most　❏ Some　❏ None

c. If yes to a, have any of your personnel been certified to any of the following levels?
(circle letters for all that apply)　　A. Awareness　B. Operational　　C. Technician

16. Wildland firefighting.

a. Is this a role your department performs?
(Check one)　❏ Yes　　❏ No

b. If yes, how many of your personnel who perform this duty have received formal training (not just on the job)?
(Check one)　❏ All　❏ Most　❏ Some　❏ None

17. Technical rescue.

a. Is this a role your department performs?
(Check one)　❏ Yes　　❏ No

b. If yes, how many of your personnel who perform this duty have received formal training (not just on the job)?
(Check one)　❏ All　❏ Most　❏ Some　❏ None

18. Basic firefighter fitness and health.

Does your department have a program to maintain basic firefighter fitness and health
(e.g., as required in NFPA 1500)?　(Check one)　❏ Yes　　❏ No

19. Infectious disease control.

Does your department have a program for infectious disease control?
(Check one)　　❏ Yes　　❏ No

PART V. FIRE PREVENTION AND CODE ENFORCEMENT

20. Which of the following programs or activities does your department conduct?
(Circle letters for all that apply)

A. Plans review　　　　B. Permit approval

C. Routine testing of active systems (e.g., fire sprinkler, detection/alarm, smoke control)

D. Free distribution of home smoke alarms　E. Juvenile firesetter program

F. School fire safety education program based on a national model curriculum

G. Other prevention program (specify)

21. Who conducts fire code inspections in your community?　(Circle letters for all that apply)

A. Full time fire department inspectors　B. In service firefighters

C. Building department　D. Separate inspection bureau

E. Other (specify)　　　　　　　　　　　　　　　　F. No one

22. Who determines that a fire was deliberately set?　(Circle letters for all that apply)

A. Fire department arson investigator　　B. Regional arson task force investigator

C. State arson investigator　　D. Incident commander or other first in fire officer

E. Police department　　F. Contract investigator　　G. Insurance investigator

H. Other (specify)

PART VI. FACILITIES, APPARATUS, AND EQUIPMENT

23. Number of fire stations:

Number over 40 years old: Number having backup power:

Number equipped for exhaust emission control (e.g., diesel exhaust extraction):

24. Number of engines/pumpers in service: (Numbers by age should sum to total.)

Total: 0 14 years old: 15 19 years old:

20 29 years old: 30 or more years old: Unknown age:

25. Number of ladders/aerials in service:

Number of buildings in community that are 4 or more stories in height: (Check one)

❑ None   ❑ 1 5   ❑ 6 10   ❑ 11 or more

26. Number of ambulances or other patient transport vehicles:

27. Portable radios. a. How many of your emergency responders on duty on a single shift can be equipped with portable radios? (Check one)

❑ All   ❑ Most   ❑ Some   ❑ None

b. How many of your portable radios are water resistant? (Check one)

❑ All   ❑ Most   ❑ Some   ❑ None   ❑ Don't know

c. How many of your portable radios are intrinsically safe in an explosive atmosphere? (Check one)

❑ All   ❑ Most   ❑ Some   ❑ None   ❑ Don't know

d. Do you have reserve portable radios equal to or greater than 10% of your in service radios? (Check one)

❑ Yes   ❑ No   ❑ Don't know

28. Self-contained breathing apparatus (SCBA). a. How many emergency responders on duty on a single shift can be equipped with SCBA? (Check one)

❑ All   ❑ Most   ❑ Some   ❑ None

b. How many of your SCBA are 10 years old or older? (Check one)

❑All   ❑ Most   ❑ Some   ❑ None   ❑ Don't know

29. Personal alert safety system (PASS) devices.

How many of your emergency responders on duty on a single shift are equipped with PASS devices? (Check one)

❑ All   ❑ Most   ❑ Some   ❑ None

30. Personal protective clothing.

a. How many of your emergency responders are equipped with personal protective clothing?

(Check one)   ❑ All   ❑ Most   ❑ Some   ❑ None

b. How much of your personal protective clothing is at least 10 years old?

(Check one)   ❑ All   ❑ Most   ❑ Some   ❑ None   ❑ Don't know

c. Do you have reserve personal protective clothing sufficient to equip 10% of your emergency responders? (Check one)   ❑ Yes   ❑ No   ❑ Don't know

PART VII. COMMUNICATIONS AND COMMUNICATIONS EQUIPMENT:

31. Multi-agency communication.

a. Can you communicate by radio on an incident scene with your federal, state, and local emergency response partners (includes frequency compatibility)?

❑ Yes   ❑ No   ❑ Don't know

b. If yes, how many of your partners can you communicate with at an incident scene?

❑ All   ❑ Most   ❑ Some

32. Map coordinate system.

a. Do you have a map coordinate system you would use to help direct your emergency response partners to specific locations?   ❑ Yes   ❑ No   ❑ Don't know

b. If yes, what system do you use? (Check one)

❑ Based on longitude/latitude

❑ Local system   Map Grid/Street Address/Box Alarm Number

❑ Based on Military Grid Reference System (MGRS)

❑ State Plane Coordinate System   ❑ Other (specify)

33. Telephone communication. Do you have 911 or similar system?   ❑ Yes, 911 basic

❑ Yes, 911 enhanced   ❑ Yes, other 3 digit system (specify)   ❑ No

34. Dispatch. a. Who has primary responsibility for dispatch operations? (Check one)

❑ Fire department   ❑ Police department   ❑ Private company

❑ Combined public safety agency   ❑ Other (specify)

b. Do you also have a backup dispatch facility?   ❑ Yes   ❑ No

35. Internet access. a.Does your department have Internet access?   ❑ Yes   ❑ No

b. If yes, describe the access you have. (Check one) ❑ All personnel have individual access

❑ One access point per station, multiple stations   ❑ One access point at the only station

❑ Access at headquarters, but there are multiple stations   ❑ Other (specify)

PART VIII. ABILITY TO HANDLE UNUSUALLY CHALLENGING INCIDENTS

Each question is based on an example incident. We want to know whether you have enough local resources to handle such an incident, and if not, how far you would have to go to obtain sufficient resources. Both the type and the size of the incident are specified to give you something specific to react to and a challenge that will often need more than local resources.

36. Technical rescue and EMS for a building with 50 occupants after structural collapse.
a. Is this type of incident within your department's scope? (Check one) ❑ Yes ❑ No
b. If yes, how far would you have to go to obtain enough people with specialized training for this incident?
(Check one) ❑ Local would be enough ❑ Regional ❑ State ❑ National
c. If yes, how far would you have to go to obtain enough specialized equipment to handle this incident?
(Check one) ❑ Local would be enough ❑ Regional ❑ State ❑ National
d. Do you have a plan for working with others on this type of incident? (Check one)
❑ Yes, written agreement ❑ Yes, informal ❑ Yes, other (specify) ❑ No

37. Hazmat and EMS for an incident involving chemical/biological agents and 10 injuries.
a. Is this type of incident within your department's scope? (Check one) ❑ Yes ❑ No
b. If yes, how far would you have to go to obtain enough people with specialized training for this incident?
(Check one) ❑ Local would be enough ❑ Regional ❑ State ❑ National
c. If yes, how far would you have to go to obtain enough specialized equipment to handle this incident?
(Check one) ❑ Local would be enough ❑ Regional ❑ State ❑ National
d. Do you have a plan for working with others on this type of incident? (Check one)
❑ Yes, written agreement ❑ Yes, informal ❑ Yes, other (specify) ❑ No

38. Wildland/urban interface fire affecting 500 acres.
a. Is this type of incident within your department's scope? (Check one) ❑ Yes ❑ No
b. If yes, how far would you have to go to obtain enough people with specialized training for this incident?
(Check one) ❑ Local would be enough ❑ Regional ❑ State ❑ National
c. If yes, how far would you have to go to obtain enough specialized equipment to handle this incident?
(Check one) ❑ Local would be enough ❑ Regional ❑ State ❑ National
d. Do you have a plan for working with others on this type of incident? (Check one)
❑ Yes, written agreement ❑ Yes, informal ❑ Yes, other (specify) ❑ No

39. Mitigation (confining, slowing, etc.) of a developing major flood.
a. Is this type of incident within your department's scope? (Check one) ❑ Yes ❑ No
b. If yes, how far would you have to go to obtain enough people with specialized training for this incident?
(Check one) ❑ Local would be enough ❑ Regional ❑ State ❑ National
c. If yes, how far would you have to go to obtain enough specialized equipment to handle this incident?
(Check one) ❑ Local would be enough ❑ Regional ❑ State ❑ National
d. Do you have a plan for working with others on this type of incident? (Check one)
❑ Yes, written agreement ❑ Yes, informal ❑ Yes, other (specify) ❑ No

PART IX. NEW AND EMERGING TECHNOLOGY

40. Thermal imaging cameras. Do you have any now or plan to acquire any?
(Check one) ❑ Now own ❑ Plan to have in 1 year ❑ Plan to have in 5 years ❑ No plan to acquire

41. Mobile data terminals. Do you have any now or plan to acquire any?
(Check one) ❑ Now own ❑ Plan to have in 1 year ❑ Plan to have in 5 years ❑ No plan to acquire

42. Advanced personnel location equipment. Do you have any now or plan to acquire any?
(Check one) ❑ Now own ❑ Plan to have in 1 year ❑ Plan to have in 5 years ❑ No plan to acquire

43. Equipment to collect chem/bio samples for analysis elsewhere. Do you have any now or plan to acquire any?
(Check one) ❑ Now own ❑ Plan to have in 1 year ❑ Plan to have in 5 years ❑ No plan to acquire

PART X. YOUR TOP 3 NEEDS IN YOUR WORDS.

44.

45.

46.